History of Techniques

SCIENCES

History of Sciences,
Field Director – Jean-Claude Dupont

Issues, General Themes and Methods of History of Sciences,
Subject Head – Stéphane Tirard

History of Techniques

Coordinated by
Guillaume Carnino
Xavier Guchet

iSTE

WILEY

First published 2026 in Great Britain and the United States by ISTE Ltd and John Wiley & Sons, Inc.

ISTE Ltd
27-37 St George's Road
London SW19 4EU
UK

www.iste.co.uk

John Wiley & Sons, Inc.
111 River Street
Hoboken, NJ 07030
USA

www.wiley.com

Library of Congress Control Number: 2025944879

British Library Cataloguing-in-Publication Data
A CIP record for this book is available from the British Library
ISBN 978-1-78945-207-5

ERC code:
SH6 The Study of the Human Past
SH6_14 History of science and technologies, environmental history

The manufacturer's authorized representative according to the EU General Product Safety Regulation is Wiley-VCH GmbH, Boschstr. 12, 69469 Weinheim, Germany, e-mail: Product_Safety@wiley.com.

Contents

Introduction

Guillaume CARNINO and Xavier GUCHET
COSTECH, Université de technologie, Compiègne, France

Technique is both a material extension of the body and an intentional projection into the tool: it is both an instrument and a skill. It is first and foremost expertise as a bodily process; then, through repetition, it gradually settles and becomes embedded in the objects produced, which is why bifaces, pencils, and screwdrivers are all considered "technical" objects[1]. Insofar as technique, as know-how, involves the mobilization of both the body and mind, it can be acquired only through experience, which alone guarantees the right balance between the intelligence of the action and the mastery of the operation. In the same way, technique as a tool (be it a pen holder or a slide rule) always involves and presupposes the joint mobilization of the body and mind. In other words, the mind informs matter, and matter informs the mind.

Actions and practices are physically embodied in the tools they use. François Sigaut has shown how the shape of sickles can be correlated with their use (Sigaut 2006). In areas where seeds are sown using a seed drill, two basic actions are performed during harvesting (grabbing a bunch and then cutting it), which results in short sickles with fully sharpened blades. In regions where sowing is performed by hand, harvesting is carried out by combining three basic movements (gathering the plants to form a bundle, grabbing the bundle and then cutting it), a sequence reflected in sickles with longer blades, part of which is not sharpened, to allow the necessary movement to gather the stems.

1 Paleoanthropology (see, for example, Boëda (2013)) teaches us how, in the very materiality of carved flint, we can perceive the trace and memory of the actions that led to its creation.

History of Techniques,
coordinated by Guillaume CARNINO and Xavier GUCHET.

Conversely, Colin Jones shows how the emergence of dentistry and odontology techniques around 1750 coincided with the development of smiling as a polite norm – first in Versailles, where aristocrats displayed their status by smiling broadly (which had previously been considered vulgar), then throughout France (Jones 2014). Here, it is the artifacts and material devices (implanted false teeth, pliers, etc.) that constitute new cultural actions/gestures and practices.

This reciprocal act of tools on actions and practices on instruments is called *transduction* by Gilbert Simondon (2017, 2020). This term refers to the existence of a metastable equilibrium (in other words, resulting from permanent and reciprocal coevolution, replayed at every moment) between technique and human beings or, more specifically, between expertise and technical objects. This transduction implies that human beings evolve their techniques as the latter redefine them, which John Culkin, commenting on Marshall McLuhan, sums up effectively: "We shape our tools and then our tools shape us" (Culkin 1967, p. 70).

In biology, adaptation is the mechanism by which function creates an organ. Under pressure from the environment, a capacity manifested by an organ will appear and stabilize because of the advantage it provides. Just as environmental constraints produce evolutionary pressure in biology – adaptation – the constraints of matter induce certain morphological invariants in technical matter: we see that, in a similar way, function also creates the organ, at least in part. Refusing to invoke divine providence, ethnologist and prehistorian André Leroi-Gourhan explains the functional similarities between objects (propellers, adzes, axes, etc.) from different points in history and culture by the concept of *tendency*, a true convergence between physiological capacities and the biomechanical and physical constraints of matter.

Exaptation is a concept initially proposed by biologists Stephen J. Gould and Elizabeth Vrba, which is the exact opposite of adaptation (Gould and Vrba 1982): it is no longer the function that creates the organ but the organ that creates the function through the functional drift of already existing structures. Tetrapods thus acquired front legs before leaving the water, which served as paddles, with fingers for grasping prey during attacks. The ability to walk, provided by the legs, appeared later, when the animals left the water (this was also made possible by an exaptation, since primitive lungs, before being used on land, were an adaptation to the oxygen-poor waters of lagoons). The mechanism is the same for feathers, which were initially used for thermal regulation and protection before being exapted for flight in certain animals. While technical adaptation occurs when the function creates the tool, exaptation occurs when the tool creates the function.

The history of techniques thus highlights numerous cycles of adaptation and exaptation. The first clocks were designed to reproduce sundials and thus emulate the principle of unequal hours, which vary according to the seasons and the length of the day: since mechanical hours are equal, attempts were made for a long time to remedy this flaw to produce unequal hours mechanically (the clock of Ctesibius of Alexandria, described by Vitruvius, had graduations that changed the length of the hours throughout the year), before the widespread use of mechanical clocks eventually completely dissociated human time from natural rhythms, creating a world regulated by mathematically identical hours (Dohrn-Van Rossum 1997). The clock, initially a simple adaptation designed to imitate sundials, embodies a new mechanical perception of time.

The first railways were used to transport coal, before it was discovered that there was a vast market for passenger transport, which ultimately completely reshaped the living environment: initially offering an alternative to river transport, the train eventually transformed the landscape and brought about contemporary industrial speed (Ollivro 2000). Similarly, the first automobiles appeared as horseless carriages; nearly half a century later, the car became the foundation on which the contemporary economy and geography are built, with the average distance between home and work now exceeding 25 km in France (Flonneau et al. 2014).

Any technique that develops on a societal scale leads to forms of exploitation. Initially developed to improve food preservation over time (for military purposes and the spread of French luxury), the tin can, once socially deployed, become a time-saving device (Fichou 2005): it became possible to "store" cooking time in these objects and then prepare a meal quickly. The tin can thus truly be the "first industrial food" (Brioist and Fichou 2012). Oil was initially used to lubricate steam engines to increase their efficiency before other engines were produced to use the power released by its combustion (Auzanneau 2015). The development of the microphone revolutionized French music in the 1930s, following in the footsteps of American crooners, by reconfiguring the vocal, gestural and facial interpretation of lyrics, public tastes, and, *in fine*, the musical repertoire (Bocquet 2021). Smartphones have created an unprecedented need for remote communication and have reduced the use of telephone conversations to a bare minimum, whereas other features that were initially secondary (such as instant messaging) have come to dominate (Simone 2012). Howard Aiken, who designed the Mark I electromechanical computer at Harvard during World War II, wrote to Jay Forrester, head of the Whirlwind project (which gave rise to the first digital computer), that "there will never be enough problems, or enough work, to keep more than one or two of these computers busy" (Edwards 2004, p. 231). It is often very difficult to imagine the exaptation that results from the widespread introduction of a technique into everyday life. Conversely, we understand that the challenge is no longer to imagine the sudden advent of a technique but rather to understand how it has gradually unfolded to the point of exaptation.

Since at least the beginning of the 21st century, historiography devoted to technique has thus abandoned the figure of the brilliant inventor in favor of a greater focus, that of the reality of incremental and progressive innovations that have punctuated technical evolution over the long course of history. Academic specialization, which is often criticized, has had profound positive effects on the history of technique over the past thirty years: by abandoning the grand narrative, historiography devoted to technique has, in many cases, also dispensed with ideology, and discussions have focused on specific points, supported by sources. In many cases, the patient's history has been completely disrupted. A canonical example, if ever there was one, is the early Middle Ages in the West, which was long perceived as a period of disintegration of the Roman Empire and barbarism. Today, it is seen as a time of political and economic structuring that gave birth to the world we live in and is now understood in all its complexity: the emergence of Christianity gave rise to a medieval European world through its growing network of bishops, the incredible complexity of political battles between "barbarian" warlords quickly discredited such pejorative labels, and the flourishing economic development of certain trades and regions formed the basis for urbanization in the early second millennium (Wickham 2005; Ruiz-Domenec 2011; Bozoky 2017). The specialization of historical writing has thus had a regenerative effect on the very content of historical narratives, and a genuine body of cumulative knowledge based on precise and controlled archival research has become widely available. This specialization has a cost (entry into the discipline is much more difficult than before) and a net benefit (the renewal and seriousness of the theses defended). Several recent collective works have thus emphasized the multiplicity of technical and industrial trajectories, thereby completing the deconstruction of the grand narrative of progress (Carnino et al. 2016; Hilaire-Pérez et al. 2016; Hilaire-Pérez and Zakharova 2016; Blond et al. 2017; Bernasconi et al. 2022; Carnino et al. 2024).

It is now difficult to measure how much of the vast majority of fields crossed by the history of techniques has been reworked, or even completely reinvented, in recent decades: from the oldest to the most recent, techniques can no longer be described as pure projections of reason, because we know that they form a material and social universe with its own dynamics and participate in the emergence of human worlds; Neolithic agriculture is no longer seen as deliverance from the burden of hunting and gathering, but as a political and hierarchical construction that ultimately laid the foundations for new forms of farming legitimacy, which continued into the Middle Ages and up to the present day; craftsmanship is no longer described as a world of routine founded by tradition, but as a world traversed by multiple collective dynamics of invention; the engineer is no longer a creative demiurge, but the linchpin of the reordering of technical hierarchies in the context of industrialization; industry is not the sudden takeoff of mechanical production, but a long process of intensification of labor, including manual labor, which continues today; technical invention is no longer central, to the point that it has sometimes become a false question; mechanization is not liberation from the hardship of work, but a domestication of social relations; the beginnings of industrialization was not an era of ecological unconsciousness, but rather a time when old environmental sensibilities were deliberately undermined in order to pollute more freely; mass production did not usher in an era of industrial abundance, but rather created a structural consumerist shortage through the use of objects designed and specialized upstream; the history of energy is not one of substitution between different sources (coal, oil, wind, etc.), but of addition; no industrial process has probably had as much material and ecological impact in history as the digital cloud today, etc. The list is too long to be exhausted here, and this book attempts, in its own way, to highlight some of the areas in which historiography has been largely rewritten over the past two decades.

The book comprises 12 chapters, including the introduction. We wanted to provide an overview of the history of techniques while meeting five requirements that we felt were important: (1) we wanted to represent several geographical areas; (2) we wanted to represent several periods equally; (3) we wanted to offer analyses of important concepts in the history of techniques and how these concepts have evolved within the discipline; (4) we also wanted to show how the history of techniques, which is very valuable for understanding the ins and outs of major contemporary challenges, has taken up some of these topics; and finally, (5) we wanted to provide an overview of the benefits that certain disciplines have gained from a closer relationship with the history of techniques.

We believe that these five initial requirements have been met, although we would have liked some of them to have been addressed in greater depth. For example, we felt it would have been interesting to include chapters on the historiography of African, South American and Oceanian techniques, among others. We also felt that it was important to include a chapter on the historiography of Antiquity. Due to the unavailability of some colleagues, we have had to scale back our ambitions somewhat. However, we believe that the book as a whole provides a faithful and relatively broad overview of the history of techniques, their current trends, their dynamism and their great richness.

I.1. Outline of the book

Using water mills as a case study, Chapter 1 revisits recent medieval historiography and debunks the persistent but false notion that Antiquity and the Middle Ages were periods of stagnation, stuck in the inertia of immutable routines, before the Renaissance kicked off the modern era. Drawing on archaeological excavations, the historiography of water mills shows that these technical devices were, on the contrary, the subject of numerous innovations, combined with the perpetuation of ancient techniques. Combining heterogeneous temporalities and multiple materialities, water mills provide access to a more detailed understanding of the social groups that used, transformed and disseminated them.

Chapter 2 provides an overview of the evolution of the historiography of techniques in China, which has gone through three different models (as was the case for other historiographies of techniques).

1) The diffusionist model, which was dominant until the mid-1990s, stipulates that inventions are the hallmark of progress and the state of advancement of a civilization. Inventions are intended to be transferred from the most advanced regions to those at a lower stage of development. The history of techniques thus appears as a linear, teleological (Eurocentric) and universalist narrative.

2) The circulation model breaks with the global conception of the diffusion of techniques from one cultural area to another, placing greater emphasis on the local and territorial nature of techniques. The circulation of techniques is not the result of a large-scale transfer process but rather of local actors.

3) The global history model recovers the global aspect of the diffusionist model but not its linear conception of technical progress: on the contrary, this global history is responsive to the complexity of re-appropriations, the plurality of actors and the disparate temporalities that affect the circulation of techniques.

Chapters 3–6 each examine a central concept (or one that has recently become central) in the history of techniques, whose reinterpretations and reworkings have accompanied the reconfigurations of the discipline itself: these are the concepts of invention (Chapter 3), material (Chapter 4), technology (Chapter 5) and industry (Chapter 6).

Thus, Chapter 3 shows that the concept of invention, which has been heavily criticized by some historians of technology on the grounds that it suggests a progressive, discontinuous and linear view of history, has experienced a strong resurgence of interest over the past 20 years. Invention is now being reinterpreted from a continuous and long-term perspective. Invention no longer means disruption or sudden progress but rather appropriation, incremental improvements, patient refinement over time, small adjustments, and local transpositions of processes and objects from one trade or sector of activity to another. Freed from its associations with a teleological and progressive conception of the history of techniques, the concept of invention is at the heart of a revival of the historiography of techniques, particularly in the 18th and 19th centuries.

Chapter 4 addresses a concept that has not occupied a central position in the history of techniques but has become a key concept in the discipline over the past 20 years: material. Not the matter of philosophers, but concrete material, extracted, transformed, mixed with other materials, from which objects are made. This emergence of material as a fully fledged and key player in the history of techniques can be partly explained by the importance of the ecological and geopolitical challenges currently posed by the materials (metals, plastics, etc.) used in the manufacture of our devices, particularly digital devices. Writing the history of techniques from the perspective of materials allows us to place the present in a longer time frame and to better understand the issues at stake. A history of techniques through the prism of materials is necessarily highly interdisciplinary: here, the historian converses with the archaeologist, the ethnologist, the art historian, the prehistorian and so on. This intertwining of knowledge and methods of analysis thus reveals materials as true historical subjects endowed with their own "agency", shaping the complex and ambivalent relationships between human societies and their environments. Finally, paying attention to materials in the history of techniques also means reopening some of the discipline's major debates: the question of innovation, the relationship between science and technology, and the relationship between techniques and culture.

Chapter 5 addresses the concept of technology. According to conventional wisdom, technology has always had a clear, unambiguous and consensual meaning in English, whereas "*technologie*" in French has two possible definitions and is therefore ambiguous: on the one hand, the word as it is currently used in everyday language translates the English word "technology" to refer to techniques that are highly dependent on the most advanced sciences (according to the conception of technology as "applied science") and linked to industrialization; on the other hand, "*technologie*" also refers, in accordance with its etymology, to a rational discourse on techniques, a science of techniques (*techne, logos*). Contrary to this preconceived idea, the chapter shows that the word *technology* is far from being devoid of ambiguity: on the contrary, it points to three meanings that are often confused. Technology translates industrial arts, the idea of applied science and the old German term *Technik* (*technique* in French). The chapter traces the tribulations of these different definitions of technology, which have strongly marked the birth of the history of techniques as a disciplinary field. The author argues that clarification is necessary to avoid misunderstandings.

Chapter 6 examines the concept of industry – a concept associated with two persistent but false ideas: on the one hand, the idea of the Industrial Revolution as a clear-cut, recent break in the history of techniques, leading to a contrast between a pre-industrial era (everything before the mid-18th century and everything outside Europe) and an industrial era; on the other hand, the idea that industrialization was a homogeneous and massive process. The chapter debunks both of these ideas. Far from being a recent and European phenomenon, industrialization characterized technical organizations in ancient times: this phenomenon must therefore be considered over a long period of time. We need to free ourselves from the discontinuist view of the history of industrialization, from notions of "disruption" and "revolution", in the sense of a sudden and global event, to identify underlying continuities. Furthermore, far from being a univocal phenomenon, industrialization took on very diverse forms: here, heterogeneity prevails. Far from appearing as a massive, homogeneous foundation, the so-called "industrial revolution" of the 18th and 19th centuries diffracts into a multiplicity of reorganizations of technical knowledge, expertise and actors, the complexity of which has been revealed by recent historiography. Does this mean that industrialization refers to a phenomenon with blurred boundaries diluted in the global historical mass, without it being possible to identify any singular characteristics of what happened in the West slightly more than 200 years ago? No, and this chapter seeks to do the opposite, to identify what makes our era unique, particularly by considering the effects of scale: when the scale changes, the nature of a phenomenon changes. While other civilizations have experienced industrial processes, our process is particularly characterized by the (global) scale on which it has developed its industry.

Chapter 7 builds on the previous chapter and focuses on the relationship between the history of techniques and the history of economics and capitalism. Historians of capitalism have long been interested in technology, with Marx being a prime example of this. However, historians of technology are still generally reluctant to incorporate the contributions of the history of capitalism into their studies. However, this observation should be qualified, as collaborations between historians of technology and capitalism have already proven fruitful on various topics, such as finance and the knowledge and skills mobilized in the workplace. Furthermore, economic historians have long questioned the relevance of a history centered on great inventors and have instead emphasized the complex processes of technique diffusion – themes that are dear to historians of techniques today. This chapter does not provide a general overview of the interactions between the history of techniques and capitalism but rather seeks to shed light on the fruitfulness of these interactions in a particular case that has been little studied: the standardization and construction of technical standards.

Chapters 8 and 9 discuss how the history of technology has taken up current social issues, in this case, ecological and gender issues.

Chapter 9 addresses the relationship between the history of techniques and environmental history. Until the 2000s, the history of techniques had not considered the environment as an essential aspect of its studies. The environment was not completely absent, but it was considered a secondary, background aspect, with a few exceptions, such as Mumford's *Technique and Civilization*, which placed the history of human societies and their techniques in their natural environments. The social turn in the history of techniques in the 1970s and 1980s further emphasized the socially constructed aspect of techniques to the detriment of consideration of natural environments. The situation changed in the 1990s, especially in the 2000s. This change should be viewed in the context of developments in environmental history itself, which has moved away from a "declinist" approach, in which humans are seen as a destructive force in nature, to a more nuanced conception of environments as intertwined heterogeneous dynamics, human activities and natural processes forming systems that are far from balanced and in constant transformation. This convergence between the history of techniques and environmental history has direct political implications: better integrating the environment into the history of techniques means being able to better understand contemporary ecological challenges by situating them in the long history of the relationship between human societies and the natural environment.

Chapter 9 addresses the relationship between the history of techniques and gender from two related perspectives: on the one hand, it examines how the social organization of technical production, reinforced by powerful masculinist discourses of legitimation, has literally produced gender inequalities in techniques, leading to the perception of women's supposed inferiority in this field as "natural"; on the other hand, it highlights the importance of women's real contribution to design, manufacturing and dissemination of techniques and to the invention of ways of using techniques throughout history. The chapter thus reveals the richness and complexity of women's relationship with techniques. Recognizing the key place women have in the technical world not only does them justice but also makes it possible to invent new technical relationships with the world based on their experiences and practices – an absolutely necessary opening to meet contemporary challenges.

Chapters 10 and 11 bring the history of technology into dialogue with other disciplines, namely anthropology and the philosophy of technology. Chapter 10 focuses on the relationship between anthropology and the history of techniques. It seeks to show how adopting a historical approach has enabled anthropologists to highlight the importance that techniques deserve in the study of "other" peoples (in other words, other than us, Westerners of the 20th and 21st centuries), leading them to abandon a notion that had long guided their view of techniques: "material culture". This notion encourages an emphasis on the symbols and aesthetic aspect of peoples' technical life, neglecting a careful examination of what is strictly technical in technology (infrastructure, production, expertise, etc.). By incorporating the contributions of the history of techniques, anthropology has thus changed the way it studies the techniques of "others" and the general meaning given to this study, from the time it was used to place civilizations on an upward trajectory of progress, with Western industrial civilization naturally at the top, to the present day, when studying the technical activities of "others" in detail gives us the means to better understand ourselves.

Chapter 11 addresses the relationship between philosophy and the history of technology. The philosophy of technology established itself as a discipline in its own right, separate from the philosophy of science, in the 1970s. However, this process went hand in hand with the opening up of this young discipline to the humanities and social sciences, particularly to new science and technology studies (STSs). The chapter examines the place of history among these disciplines, which have contributed greatly to the philosophy of technology. The various seminars that have attempted to bring philosophy and the history of technology into dialogue from the 1970s to the 2000s show that this dialogue took time to take shape and pay off. Finally, it offers a general overview of existing collaborations between philosophy and the history of technology, detailing the four perspectives through which these collaborations take shape: individual trajectories, traditions of thought (some of which are much more inclined than others to intertwine philosophy and history in their study of technology), shared concepts and the historicization of concepts and issues in the philosophy of technology.

I.2. References

Auzanneau, M. (2015). *Or noir. La grande histoire du pétrole*. La Découverte, Paris.

Bernasconi, G., Carnino, G., Hilaire-Pérez, L., Raveux, O. (eds) (2022). *Les réparations dans l'histoire. Cultures techniques et savoir-faire dans la longue durée*. Presses des Mines, Paris.

Blond, S., Hilaire-Pérez, L., Virol, M. (eds) (2017). *Mobilités d'ingénieurs en Europe, XVᵉ-XVIIIᵉ siècle*. PUR, Rennes.

Bocquet, T. (2021). *Le Charme du microphone. Métamorphoses de la chanson dans la France des années 1930*. UGA, Grenoble.

Boëda, É. (2013). *Techno-logique et technologie. Une paléo-histoire des objets lithiques tranchants*. Archéo-éditions, Bordeaux.

Bozoky, E. (ed.) (2017). *Les saints face aux barbares au haut Moyen Âge. Réalités et légendes*. PUR, Rennes.

Brioist, P. and Fichou, J.-C. (2012). La sardine à l'huile ou le premier aliment industriel. *Annales de Bretagne et des Pays de l'Ouest*, 119(4), 69–80.

Carnino, G., Hilaire-Pérez, L., Kobiljski, A. (eds) (2016). *Histoire des techniques. Mondes, sociétés, cultures, XVIᵉ-XVIIIᵉ siècles*. PUF, Paris.

Carnino, G., Hilaire-Pérez, L., Lamy, J. (eds) (2024). *Global History of Techniques: 19th-21st Centuries*. Brepols, Turnhout.

Culkin, J.M. (1967). A schoolman's guide to Marshall McLuhan. *Saturday Review*, 70–72, 51–53.

Dohrn-van Rossum, G. (1997). *L'histoire de l'heure. L'horlogerie et l'organisation moderne du temps*. MSH, Paris.

Edwards, P.N. (2004). Construire le monde clos : l'ordinateur, la bombe et le discours politique de la guerre froide. In *Les sciences pour la guerre, 1940-1960*, Pestre, D. and Dahan, A. (eds). EHESS, Paris.

Fichou, J.-C. (2005). La grande guerre et les conserveurs de sardines. *Guerres mondiales et conflits contemporains*, 219(3), 71–86.

Flonneau, M., Laborie, L., Passalacqua, A. (eds) (2014). *Les transports de la démocratie. Approche historique des enjeux politiques de la mobilité*. PUR, Rennes.

Gould, S.J. and Vrba, E. (1982). Exaptation, a missing term in the science of form. *Paleobiology*, 8, 4–15.

Hilaire-Pérez, L. and Zakharova, L. (eds) (2016). *Les techniques et la globalisation au XXᵉ siècle*. PUR, Rennes.

Hilaire-Pérez, L., Simon, F., Thébaud-Sorger, M. (eds) (2016). *L'Europe des sciences et des techniques. Un dialogue des savoirs, XVᵉ-XVIIIᵉ siècle*. PUR, Rennes.

Jones, C. (2014). *The Smile Revolution in Eighteenth Century Paris*. Oxford University Press, Oxford.

Ollivro, J. (2000). *L'homme à toutes vitesses : de la lenteur homogène à la rapidité différenciée*. PUR, Rennes.

Ruiz-Domenec, J.E. (2011). *Entre historias de la Edad Media*. Editorial Universidad De Granada, Grenada.

Sigaut, F. (2006). Le savoir des couteaux. Dire le savoir-faire. *Cahiers d'Anthropologie sociale*, 1, 133–139.

Simondon, G. (2017). *On the Mode of Existence of Technical Objects*. Univocal Publishing, Minneapolis.

Simondon, G. (2020). *Individuation in the Light of Notions of Form and Information*. University of Minnesota Press, Minneapolis.

Simone, R. (2012). *Pris dans la toile. L'esprit aux temps du Web*. Gallimard, Paris.

Wickham, C. (2005). *Framing the Early Middle Ages: Europe and the Mediterranean, 400–800*. Oxford University Press, Oxford.

1

The Medieval Watermill: A History of Technology

Lisa CALISTE[1] and Catherine VERNA[2]

[1] *Inventaire général du patrimoine culturel, Montpellier, France*

[2] *ARSCAN, CNRS, UMR 7041, Université Paris 8 Vincennes-Saint-Denis, Paris, France*

1.1. Introduction

The mill, considered the innovation that led to the expansion of mechanization in Western civilizations, is a prime subject for historiographical research in the history of technology. The article published by Marc Bloch in 1935 in a special issue of the *Annales* journal, which was devoted to the history of technology, serves as a starting point, as it established the direction of research for several decades, from the question of the origins and spread of innovation to the analysis of social structures as drivers of technical change (Bloch 1935a[1]). Returning to this often-cited text proves enriching, mainly through the comparison between the original demonstration and the reinterpretations of historians. The practice is limited here to the watermill, excluding the windmill, and focuses on French historiography. Academic work, seminars and collective research programs form a vast and solid basis for reviewing historiography since the 1930s and establishing the current state of research. Since the 1980s, data have multiplied, particularly with respect to regional monographs and archaeology. The following decade was marked by two major interdisciplinary meetings: the conference of the *Société d'archéologie médiévale* (French Medieval Archaeology Society) on the theme of technical innovation (Beck 1998) and the *Journées d'histoire de l'abbaye de Flaran* (Flaran Abbey History Days), which was

1 For the context of the article, see Arnoux (2012, pp. 294–301).

History of Techniques,
coordinated by Guillaume CARNINO and Xavier GUCHET. © ISTE Ltd 2026.

devoted to mills and millers in the European countryside (Mousnier 2002). The seminar held in Lons-le-Saunier in 2011 offered a unique overview of the archaeology of watermills (Jaccottey and Rollier 2016a). The aim is therefore to compare Marc Bloch's text with the sources that have been unearthed and the methods developed by medievalists and to propose a summary of current knowledge based on the themes of the 1935 article, namely, the question of innovation and technical temporalities, the materiality of mills, and, finally, the study of groups and individuals as drivers of technical change.

1.2. The era of mills

The mill has been the subject of numerous investigations with regard to its origins, which are based on texts, iconography and, more recently, material remains. Very early on – Marc Bloch's article is a case in point – historians adopted a long-term thinking approach, highlighting phases linked to the "invention"[2], innovations and diffusion of hydraulic techniques in ancient, medieval and modern societies. An early distinction was also made between grain mills and industrial mills, such as fulling mills, bark mills, paper mills and hydraulic forges. Since the publication of Marc Bloch's article, research has focused on the creation period of hydraulic machines, in all their typological diversity, combined with reflection on the diffusion of innovation and technical flows.

1.2.1. *In search of origins*

Already present in the writings of Viollet-le-Duc[3], the search for the origins of the mill was revived by Marc Bloch in 1935. His article, which gave pride of place to the ancient period, opened the debate on the use of hydraulic energy in the first centuries of our era. His famous phrase, "an ancient invention, the watermill is medieval because of its era of true expansion", provided fruitful ground for the theory of the "blockage of techniques" during antiquity, a theory that emerged in the 1930s. The arguments put forward by Marc Bloch to explain this blockage,

2 Marc Bloch used the term "invention", while acknowledging the limitations of this "oversimple word" (Bloch 1935b). At the same time, Schumpeter (1935) also used the term "invention", although he was undeniably describing what he would later call "innovation". The latter, and the concept associated with it, were in fact only developed and disseminated from the late 1930s onward.

3 In the mid-19th century, in his *Dictionnaire raisonné de l'architecture française du XIe au XVIe siècle*, Viollet-le-Duc published the dates and contexts of the appearance of water, wind and tide mills.

including the use of numerous slaves, were taken up and developed after the war, particularly in the 1950s and 1960s. During this period, historians such as André Aymard, a Hellenist, proposed a cultural explanation, namely, the devaluation of productive activities among the elite. It was against this backdrop that Moses I. Finley imposed the idea of the stagnation of techniques during Antiquity, ruling out the possibility of massive recourse to hydraulic energy and, even further, the existence of an ancient industry (Pagès and Verna 2022). As early as 1999, at the *Journées de Flaran* (Flaran Days) devoted to mills and millers, medievalists echoed the work of archaeologists, who questioned the "blockage of techniques" attributed to Antiquity, notably by renewing our knowledge of the spread of the watermill in Roman times (Comet 2002, pp. 10–11). The abandonment of Marc Bloch's model, which was based on limited written sources, became clear less than a decade later, owing to excavations of ancient watermills, which have been steadily increasing in recent years (Wikander 2000; Jaccottey and Rollier 2016a; Buchsenschutz et al. 2020; Wilson 2020). We now know that the watermill, attested in both rural and urban environments, spread rapidly in Narbonnaise, inland Gaules and Germania during the first two centuries of the Roman Empire. The use of water power was commonplace in the *villae* of the Narbonnaise region to grind grain and produce the flour needed to feed the area's workforce (Brun 2016).

The search for origins, which was one of Marc Bloch's major preoccupations, led him to take an interest in the emergence of new hydraulic machines, both agricultural and industrial, whether they were olive presses, bark mills, fulling mills or "martinet" (trip hammers), as he called them (Bloch 1935a, p. 543). It was not until the mid-1950s that this line of research was taken up by the technical historian Bertrand Gille (Gille 1954). In a first essay entitled *Le moulin à eau, une révolution technique médiévale* (The watermill, a medieval technical revolution) published in 1954, he based his approach on both written sources, which he was experienced in as a chartist, and iconographic sources. In 1960, as a specialist in modern metallurgy, he published *Les origines du moulin à fer* in the *Revue d'Histoire de la Sidérurgie* (Gille 1960). The same year saw the publication of Anne-Marie Bautier's article entitled *Les plus anciennes mentions de moulins hydrauliques industriels et de moulins à vent* (The earliest references to industrial watermills and windmills) (Bautier 1960). The approach taken by historians, which was standard in the history of technology at the time, was to identify the advent of a term designating a different way of doing things, a new tool or a new machine (Bernardi and Verna 2001, p. 37). Interest was revived in the 1980s and, even more so, in the 1990s, when medievalists adopted the concept of innovation, as stated by the economist Joseph Aloïs Schumpeter. Paul Benoît, in particular, proposes a possible reading of the concept, only retaining aspects of the definition that he believes are applicable to the Middle Ages and emphasizing the radical disruption that constitutes innovation, corresponding to the creation of a new good and/or a new method of production (Benoît 1998).

While the use of hydraulic power for non-agricultural production activities – marble sawing, ore and bark grinding – is attested as far back as Antiquity, historians agree on the multiplication and spread of industrial mills in the Middle Ages, particularly during the period from the 10th to 13th centuries. These four centuries were marked by the diversification of uses for water power associated with cams and by the multiplication of industrial hydraulic sites. Fulling mills first appeared in Italy in the 10th century and are attested in Dauphiné in the middle of the 11th century and in Forez and Normandy at the end of that century[4]. Hemp mills, mentioned in written sources from the 11th century onward, were established in Moselle a century and a half earlier, as evidenced by the remains unearthed at Audun-le-Tiche (Rohmer et al. 2016). Bark mills, used to crush bark for hide processing, were present throughout much of the kingdom of France as early as the 12th century. The hydraulic sawmill, reported since Antiquity, is again attested in Ariège at the end of the 13th century (Verna 2001). Paper mills, probably the first of their kind, were present in Troyes in the second quarter of the 14th century. However, the respective chronologies of these various industrial mills – their appearance and their spread – remain difficult to establish in the absence of sufficiently numerous and in-depth archival and archaeological studies. Fewer than a dozen textile mills have been excavated in France, including the fulling mills at Pennedepie (Calvados), Bourges (Cher), Billezois (Allier) and the hemp and flax grinding and threshing mill (scutching mill) at Audun-le-Tiche (Moselle).

In this context, the knowledge concerning medieval metallurgy and its use of hydraulic power is an exception[5]. The mechanization of ironworks in the Middle Ages has been the subject of numerous studies, some of which are not devoid of nationalist political issues, as demonstrated by Catherine Verna's investigation of the hydraulic reduction forge in Catalonia (*la farga catalana*) (Verna 2008, 2017a, pp. 197–215). The idea that Catalonia was the birthplace of hydraulic power application for forging, at least as early as the 12th century, stems from a historiography centered on a technical object with a strong identity. The research focused in particular on innovation in hammering. The approach consisted of identifying the words and expressions designating a new way of doing things, such as the "*martinets*" (trip hammers) of Dauphiné and Savoy in the early 14th century (Belhoste 1992) or the "*forges renardières*" of Normandy in the 15th century

4 For information on the fulling mill, see Carus-Wilson (1941), Cardon (1996) and Lucas (2016).

5 More specifically, on millstones for crushing ore and slag, see Benoît (1997), Verna (2017a) and Bailly-Maître (2019).

(Belhoste et al. 1991; Arnoux 1993), and, more specifically, identifying the innovation by describing it in its environment – a "technical territory" (Verna 2001). This approach has been fruitful in the field of Cistercian ironmaking (in Champagne and Burgundy) and in southern areas corresponding to the technical territory of the "*mouline*" (*molina ferrea*) (a hydraulic forge) (Verna 1995, 2001, 2017a). It has benefited from the contributions of archaeology thanks to field surveys, including the study of the industrial buildings of Fontenay Abbey[6] (Côte-d'Or) (Benoît and Cailleaux 1991) and the excavation of the metallurgical installations at Castel-Minier (Ariège) (Téreygeol, forthcoming), and thanks to the analysis of products derived from metallurgical activities (Dillmann and L'Héritier 2016). For example, Cistercian abbeys in Champagne and Burgundy have been shown to be pioneers, using the "iron mill" (a direct reduction forge equipped with a hydraulic hammer) as early as 1135 (Verna 1995; Benoît and Berthier 1998). In the southern part of the French kingdom, the first known mouline – a direct reduction forge equipped with a hydraulic hammer – is that of Escoussens (Tarn), attested as early as 1283 (Verna 2001). The mouline spread rapidly throughout Languedoc and the Pyrenees: it was present in Vicdessos as early as 1299, Fenouillèdes in 1311 and Vallespir in 1314. Although, in its initial configuration, the mouline may have been equipped with water-driven bellows, the latter were undoubtedly present in 1416 in the mouline de Léca (Corsavy, Pyrénées-Orientales) (Verna 2017a). Similarly, the origins of the Walloon finery process, another method of making iron and steel by decarburizing cast iron, have been investigated both in texts and in the field (Glinet forge, Seine-Maritime) (Arribet-Deroin 1998). The first blast furnaces required hydraulic energy to increase the temperature and reach the melting point (cast iron); this same energy activated the decarburization process in the refinery. The first indications of indirect reduction, regardless of the process used, are attested in the second half of the 13th century, in several regions of northwestern Europe (Rhineland, Sweden), then in the kingdom of France from the end of the 14th century (Pays d'Othe) and the beginning of the 15th century (Cher, Côte-d'Or) (Belhoste et al. 1991, 1994). In conclusion, while the search for the origins of the watermill has helped diversify thinking about the uses of hydraulic machines, it has also led to a lasting bias in research by neglecting the study of preexisting tools and machines, particularly the blood mill (animal-driven mill), whose history largely eludes medievalists (Jarrige 2023).

1.2.2. *Technical temporalities*

Marc Bloch comments at length on the permanence of technology – milling processes using human or animal power. To explain the continued use of technology

6 The forge of the Fontenay Abbey was excavated in 2008 (Rouzeau 2020).

that predates the hydraulic mill, he emphasized the advantages offered by the hand millstone: it could be used in times of siege, was easily transported, more readily acquired and not subject to seigneurial rights (Bloch 1935a, pp. 548–552). However, by inscribing the continuum of uses in the narrative of innovations and the advent of machinism, he only sees "rural routines" or the "obscure prestiges of tradition" (Bloch 1935a, p. 550, 556, 560) in these "perpetual returns to old and rudimentary technology". Furthermore, by conceiving a "war of water and wind against the strength of arms" (Bloch 1935a, p. 557), he initiates what will later be interpreted as the rejection of novelty, that of the machine, particularly by workers, although this argument only underlies his demonstration. While resistance to mechanization has been identified by certain medievalists, particularly Jean-Louis Roch, who has provided a social and political explanation for it (Roch 2013, p. 27–29), it was affirmed as a cultural trait of medieval societies by Jacques Le Goff (1964, p. 254), commented on by Verna (2017a, p. 20). In line with Marc Bloch's text, the long-lasting use of mortars and blood and hand millstones has therefore long been perceived as an archaism, a manifestation of a supposed medieval immobility in production techniques. However, as early as 1965, this view was challenged by Marxist historian Charles Parain, who, according to the ethnological approach he defended, emphasized the "detail improvements" to be considered in the study of tools (Parain 1965, p. 58).

Research carried out over the past generation, particularly by modernists and contemporaneists such as Serge Benoît (2020), has enabled us to rethink the rhythms and periodizations of techniques and to highlight the plurality of technical temporalities[7]. Moreover, by exploring the field of "routine" and everyday life, on the scale of men and workplaces, they have emphasized the virtues of repetition as a source of technical mastery (Woronoff 1998). Medievalists were quick to seize upon these tools for reflection and, following a continuist approach, also asserted the durability of ancient techniques due to their efficiency (Verna 2001). There was undoubtedly a plural temporality to milling processes based on the coexistence of novelties with old techniques validated by usage. This was also true for industrial activities, which may or may not have used hydraulic power, depending on the products expected and the capital available. The use of fulling mills, for example, did not rule out manual fulling methods, with hands and feet, for different products (Caliste 2021). Similarly, the advent of hydraulic hammering did not rule out the continued use of foot-operated forges for several decades (Verna 2017a, pp. 170–171). An addition to this is the recent rereading of the concept of innovation

7 Please refer to the presentation of the "*Techniques et temporalités*" workshop-concept, organized on October 12, 2022, as part of the GDR2092 TPH "Techniques et production dans l'histoire" coordinated by Liliane Hilaire-Pérez, Guillaume Carnino and Catherine Verna.

by historians of ancient periods. Innovation in the Middle Ages has been recognized by historians, and it has been shown that innovative breakthroughs can sometimes result from minor adjustments made in the workshop (Braunstein 1998; Verna 2000). These innovation processes are rarely grasped by the medievalist, who only sees them emerge at their culmination, when the novelty is finally recorded in writing and gives the illusion of the innovation's emergence (Verna 2017a). In the same way, archaeological remains reveal only one state of innovation, often linked to the abandonment of the production site.

The position of the wheel has crystallized thoughts on innovations in hydraulic power. For a long time, the debate was confined to a comparison between horizontal and vertical wheel mills in an attempt to establish the chronology of their respective appearance and spread. While these approaches may seem outdated today, it is interesting to look back at the original hypothesis – the primacy of the vertical wheel over the horizontal – and the steps taken to overcome this assumption. As early as 1935, Marc Bloch sketched out the hypothesis of a later appearance of the horizontal wheel, describing it as a "technical regression" from the wheel described by Vitruvius (Bloch 1935a, p. 544). This theory, which the author called for further study on, was subsequently taken up and by the late 1950s, the use of the horizontal wheel during the first centuries of our era had been proven (Moritz 1958). Today, antiquarians propose the following sequence, which, while not unanimous, is widely shared: the horizontal wheel would appear after the adoption of the vertical wheel driven by the weight of the water (overshot wheel), then the vertical wheel driven by a current (undershot wheel) (Spain 2008, quoted by Brun 2016, p. 39). With respect to the spread of innovation, already present in Marc Bloch's work, it was long limited to the Mediterranean world as the home of the watermill and the center of its influence (Bloch 1935a, p. 540). The interest shown in horizontal wheels has made it possible to question a reductive "center-periphery" pattern and to take into account technical flows, thus participating in the reflection on the complex and nonlinear diffusion of technical knowledge (Hilaire-Pérez and Verna 2016). Following the archaeological discoveries of the last 30 years, it now seems accepted that early medieval mills, which were already present in large numbers in Carolingian times (Bonnassie 1975; Champion 1996; Durand 2002), are descended from two lineages, one transmitted by the Roman world, favoring the vertical wheel, and the other diffused from a Scandinavian focus, making use of the horizontal wheel since the first-century CE (Benoît 2016). These focal points are associated with areas of distribution: from the 6th to 10th century – a time when industrial mills were rare – the vertical wheel was predominantly used in an area that included Gaul, the southern part of Great Britain and western Germania, whereas the horizontal wheel dominated in Scandinavia and a large part of the British Isles. This distribution pattern does not, however, account for the massive presence of horizontal wheel mills in Languedoc and Catalonia from the Carolingian era onward (Phalip 1992; Bolòs 2002). In addition, archaeological work has shown that the horizontal wheel was still in use in Ireland in the 12th and 13th centuries, invalidating the theory of chronological succession from the horizontal wheel to the vertical wheel from the 11th century onward. The concurrence of techniques is now recognized: the adoption of the horizontal wheel or the vertical wheel responds to different contexts in terms of hydromorphology, uses and costs. While the vertical wheel is not universal (Amouric 1980, 1984; Comet 2002), historians agree on the development of more powerful facilities, using the vertical wheel from the 12th century onward, with higher capacities for milling and corresponding to industrial uses, particularly in textiles and metallurgy.

1.3. What is a mill?

In 1935, when the history of technology was still in its infancy, the expectations of the historians of the *Annales* school were forcefully stated: the aim was to understand, describe and reconstruct ancient tools using an interdisciplinary approach, associating historians with what Lucien Febvre called “technicians”, who practiced technology, as well as archaeologists, ethnologists and linguists (Bloch 1935b; Febvre 1935). Similarly, Charles Parain (1965) advocates “going into sufficient detail” about equipment, a *sine qua non* for understanding the techniques at work in a society and for writing the history of technology. However, despite the voluntarist approaches of historians, influenced by anthropologists, the corpus of texts, images and artifacts remained insufficient for a long time to explore the materiality of medieval mills, as witnessed by the major sagas on the history of technology led by Maurice Daumas and Bertrand Gille in the 1960s and 1970s (Daumas 1962, 1965; Gilles 1978). From the 1980s onward, a renewed interest in material culture – objects and their history – led to responses that were closer to the research incentives of previous generations, particularly by implementing interdisciplinarity (Verna and Dillmann 2018). Scientific documentation has been growing for several decades, and we certainly owe this to preventive archaeology but also to the work of (re)reading archives, in particular seigneurial accounts and notary registers. The methods applied to the collected corpus – those of source criticism, which requires a meticulous investigation of words; those of archaeology and archaeometry – now make it possible to achieve precise knowledge of hydraulic equipment.

1.3.1. *Hydraulic developments and structures*

At a time when the restoration of the ecological continuity of waterways – as called for by the European Union – threatens the destruction of man-made structures, particularly dams and causeways, it is a shame that no summary has been published on structures linked to the diversion and channeling of water to harness its energy. Historians with a strong ethnological approach were the first to take an interest in these issues. As early as the 1960s, Charles Parain (1965, p. 65) emphasized the scale of the work required to divert rivers, adding to Marc Bloch's study, which made no mention of the matter. Twenty years later, François Sigaut (1986) took up the same theme, establishing a link between the spread of fish ponds and the expansion of mills in the countryside during the 10th and 11th centuries, indicating the mastery of new techniques, such as earthworks and water control, which were conducive to the development of watermills. However, historians have turned their attention away from water channeling to the location of mills, an area that has attracted the most attention, starting with Bertrand Gille's first inventory in 1954 (Gille 1954). Gille distinguished between run-of-river mills installed on rivers with large, constant flows, diversion mills adapted to streams with abundant or rapid flows (which could damage the mill), and pond mills built on streams with low flows. He also mentioned the existence of mills on high-flow streams, also known as "ship mills", as well as mills installed on bridges, the main models of which were described by Georges Comet in 1992 (Comet 1992, pp. 429–434). However, archaeological findings have enriched this classification, at least for the earliest periods. Excavations at Audun-le-Tiche (Moselle) (Rohmer et al. 2016) and Thervay (Jura) (Rollier et al. 2016) prove that early medieval mills, small wooden installations, changed location according to the river's divagations. It was only in the course of the 11th century that permanent mills were established in the Mâconnais and Clunisois regions and that rivers were developed in such a way as to channel part of their flow to the wheels in a sustainable manner (Rollier 1998; Jaccottey and Rollier 2016b, p. 845).

The construction of levees across a waterway to retain and control part of the flow requires technical skills that were of later interest to medievalists. However, Sylvie Caucanas, in her book on mills and irrigation in Roussillon, devotes a chapter to dams, dikes and channels (Caucanas 1995, pp. 123–140), which goes beyond Germain Sicard's brief description of Toulouse mills (Sicard 1953). The dam (*paxeria*, *resclausa*), which diverts water from a river into a channel, could be made out of wood, with beams placed horizontally between the rocks of an incised riverbed, or out of wood and masonry, often in environments of soft ground and wider waterways. In the latter case, two techniques have been used. The first consists of driving in rows of pilings linked by transverse beams or fascines (an assembly of branches); the squares thus form and are then filled with stones and pebbles. The second technique requires the use of trestles sunk deep into the sand. For the construction of these hydraulic structures, materials are taken from nearby sites.

Following the levee, the feeder channel (*bief, béal, rech*) carries the water to the mill wheel, and also serves to increase usable power by increasing the gradient above the wheel, thus artificially increasing the fall height. The shapes and materials of channels vary, from simple trenches dug in the ground, sometimes paved or reinforced with wooden planks, to underground conduits and aerial channels (Caucanas 1995, pp. 128–135). Their dimensions, linked to the relief, the flow of the waterway and the power needed, also vary from one geographical context to another, as well as from one era to another. Records from the 13th and 14th centuries in the Roussillon region indicate widths between 1.50 and 2 m, whereas depths remain unknown. Archaeologists, meanwhile, have uncovered 5 m-wide reaches at milling sites in the northern half of France. Channeled water may be directed as close as possible to the wheel by a wooden runner (Colomby, Thervay, Bourges) or, in the case of horizontal wheels, by a stone chute or wooden pipe. By accelerating the effective speed of the jet and avoiding lateral losses, the addition of these ducts improves the efficiency of the hydraulic motor. The reservoir, a basin used to store water upstream of the mill and sometimes used for fish farming (pond-reservoir), can also be used to artificially increase flow, to power the wheels in the event of drought or to protect them from damage caused by rainstorms (Caucanas 1995, pp. 151–152; Benoît and Berthier 1998). Power can be improved by accelerating and regulating the water velocity and, consequently, the wheel rotation speed by means of a sluice gate that is placed immediately upstream of the wheel (Belle Église, Saleux).

Generally speaking, because of the value of hydraulic structures and their vulnerability to flooding, particular care is given to them: medieval acts repeatedly state that the dam must be kept in good condition and repaired if necessary, and the channels should be cleaned regularly.

1.3.2. *Horizontal wheels, vertical wheels*

As mentioned above, historians, such as Marc Bloch, first focused on the water wheel and, more specifically, on its position, which gave rise to the distinction between vertical and horizontal wheel watermills. Depending on the movement and application needed, the vertical wheel mill is equipped with a right-angle gear (vertical gear and lantern) or a camshaft, whereas the horizontal wheel turns a vertical shaft that transmits rotation directly to the millstone. While vertical and horizontal wheel mills may have been in use simultaneously and in the same geographical area, as was the case in certain southern regions until the 19th century (Rivals 1984), the former are now much better known, benefiting from a long-standing and enduring interest. For horizontal wheel mills, owing to the more tenuous remains, reduced iconography and rarer investigations, it is difficult to extend beyond the typology outlined by Georges Comet in 1992. Comet identified two main models: the flying waterwheel and the turbine, with the wheel enclosed in a fitted masonry box (Comet 1992, pp. 440–441).

The typology of vertical wheels was established according to the position of the water inlet, a point already made by Marc Bloch when he mentioned a wheel "driven from below" and another model in which the water falls on the paddles "toward the top" (Bloch 1935a, p. 544). Medieval iconography, analyzed by Perrine Mane, depicts these two types of wheel, the top-fed wheel (also known as the "overshot") and the bottom-fed wheel (also known as the "undershot") (Mane 2002). To the previous two, we add the breastshot wheel, which receives water through a chute between its top and axle, and the side wheel, whose lower part is hit by the water. These wheels probably existed as early as the Middle Ages, owing to the mastery of the techniques required to set them in motion. The choice between these different types depends not only on local hydrological conditions but also on the energy required for agricultural and/or industrial needs and, consequently, on the capital available, as the construction and maintenance of the mechanisms are costly – we will come back to this later (section 1.4.1). The top-fed wheel is equipped with buckets (or troughs), which, when filled with water, turn the wheel under their own weight. This type of wheel is best suited for waterfalls over 3 m high and to low-flow streams (Comet 1992, p. 436). Mills equipped with an overshot wheel are located either on a bypass or, where there is a sufficient break in the gradient, on the water's edge, in which case the water is conveyed by a runner from the dam. Undershot wheels, on the other hand, are equipped with flat, sometimes curved blades or vanes on which the water pushes by percussion and/or simple entrainment. In this case, it is the flow of the waterway that drives the impeller. This model is preferably used on plains or regions with moderate relief and for medium-sized rivers (Comet 1992, p. 434). In this case, the mill is installed on a channel, often of smaller dimensions, or immediately downstream of the water reservoir.

The work of archaeologists has enabled us to identify the evolution of undershot wheels – the only ones excavated in France, if we stick to the findings of the 2011 symposium held in Lons-le-Saunier. Let us not forget that undershot wheels are also the most common type of wheel in medieval iconography. At the beginning of 2010, the archaeological corpus comprised eighteen hydraulic installations spread over fifteen excavated sites, mostly concentrated in the northern half of France. Eight mills were in operation between the 7th and 10th centuries, three between the 11th and late 12th centuries, and seven between the 13th and 15th centuries. On the basis of archaeological data, Pierre Mille proposed an original basis of thought on the power provided by mills from the 7th to 14th century (Mille 2016). He highlighted the use of wheels with monoxyl blades (Audun-le-Tiche, Thervay) during the High Middle Ages, whose limited power seems to have been reserved for milling. They were replaced by mills with assembled paddles – wider than the monoxyl-blade ones – which came into widespread use from the 11th century onward, becoming the norm in the last centuries of the Middle Ages. This record answers a question raised by Bertrand Gille in 1954 concerning the orientation of the blades: at Notre-Dame-de-Marillais (Maine-et-Loire), Colomby (Manche) and Thervay (Jura), the blades were already inclined to facilitate penetration in the water, attesting to a desire to optimize hydraulic installations. These large waterwheels made it possible to increase mill productivity by implementing increasingly voluminous equipment, particularly ever-larger millstones. By the end of the Middle Ages, sources confirmed the increase in the size of undershot wheels (diameter and width) and, consequently, in the mills' energy performance (Marchandin 2021).

Records, particularly those of the abbeys and cathedral chapters of the northern half of France, which have been preserved for the 14th and 15th centuries, provide detailed descriptions of the wheels, from their materials to their assembly, making it possible to go beyond the simple mention of a wooden wheel strapped with iron. For example, the vertical wheels on the mills in the Sens region were made from naturally curved wood, most often oak, selected from forests close to the mill sites and quickly processed after felling (Rouillard 2016). The curved elements are connected to the motor axle by several arms, the whole being assembled using dowels and hundreds of nails. Iron links and hoops reinforce the wheels as well as the gears. There are numerous metal parts in a gristmill: in addition to the reinforcements, the trunnions are also made of iron; the spindle, the bearing and the millrind are made of steel (Rouillard 2011). The wheels are set in motion by the force of the water, thanks to paddles – around 30 in the case of the Sénonais wheel from the early 15th century and 28 in the case of Toulouse wheels from the same period (Sicard 1953). For the last centuries of the Middle Ages, historians also have meticulous descriptions of mill mechanisms, including the wheel, recorded in inventories (charges, states, valuations) drawn up for each lease change and preserved in notary registers. These texts also describe the furnishings and tools found in the mills.

1.3.3. *Tools in motion*

Marc Bloch's interest in the machinery and equipment of the watermill is limited to the motor. He does not mention the chamber of millstones, let alone the hydraulic hammer. Some of the most significant advances made since the publication of the issue of *Annales* undoubtedly concern our understanding of the work spaces within mills, particularly the equipment they contain. First and foremost, particular attention has been given to power transmission mechanisms. This is mainly the gear mechanism (vertical gear and lantern), whose dimensions evolved over the course of the Middle Ages, even if they have yet to be precisely defined, as well as the camshaft. The use of the camshaft, which transforms continuous rotary motion into vertical reciprocating motion, allows for diverse industrial applications: hydraulic power is used for fulling, hammering and/or blowing for iron production, bark crushing for hide tanning, shredding rags for paper manufacture and wood sawing. In addition to this often repeated list of applications, two areas have benefited from a multidisciplinary approach: gristmilling and hydraulic forging. While these two fields of study do not address the same historiographical issues, they do share some common features in terms of the cross-fertilization of disciplines and corpuses to realize the machine.

The general physiognomy of the flour mill is difficult to reconstruct, as too few studies have been carried out on the architecture of a facility that is undoubtedly marked by the multiplicity of its forms in time and space and by the perennial nature of its location. Early medieval mills in the northern half of France appear to have been modest in size, sometimes measuring 2 m by 2 m, and built of wood on supporting posts (Audun-le-Tiche, Thervay). Mills on stilts were also present in medieval towns, including Paris, before the 13th century, as shown by a recent enlightened re-reading of the *Life of Saint Denis* (Marchandin 2021). In contrast, the mills of southern France, in the Hérault, Orb and Vidourle basins, in use during the 12th and the 13th centuries, are masonry constructions with vast footprints, sometimes several storeys high (Phalip 1992). These buildings housed part of the mill's machinery in a gear pit for vertical-wheel mills or a water chamber for horizontal-wheel mills. However, let us not forget that the milling room is at the heart of the machine and that hydraulic mechanics are used to drive the runner stone. Following the work undertaken for the ancient period (Longepierre 2012; Buchsenschutz et al. 2020), medieval millstones, which have long been neglected by historians, have been the subject of recent research programs aimed at compiling a corpus of artifacts and archival material related to milling tools from the 5th to 16th centuries[8]. A first synthesis, presented in 2011, established a state of knowledge on

8 Collective research program "Groupe Meule", initiated in 2008.

the materials making up the millstone, their provenance and their typological and morphological evolution on the basis of a corpus of 233 millstones from 54 different sites (Belmont et al. 2016). The survey is continuing, with the inventory carried out in Champagne-Ardenne having added 19 millstones, mostly dating from the 12th century, to the corpus of medieval milling tools (Minvielle-Larousse et al. 2017). In terms of supply, local extraction was the norm, with millstones transported no more than 50 km for the period prior to the 10th century. This distance increased during the late Middle Ages, marked by the beginning of the spread of millstones from the Paris Basin. With respect to millstone dimensions, historians agree that the diameter of millstones increased steadily during the Middle Ages, in line with the development of hydraulic techniques. While the diameter of millstones used in the 5th century was between 50 and 80 cm, that of Carolingian millstones was between 60 and 80 cm, with the meter being exceeded from the 10th and 11th centuries. A steady rise in size is thus observed between the 7th and 12th centuries, followed by a rapid increase in the 13th to 14th centuries, reaching 120–150 cm by the end of the Middle Ages. In the 15th century, such dimensions are attested both in and around Paris (Marchandin 2021) and in the southern countryside (Caliste 2021).

The second record, devoted to hydraulic forging and the mechanization of ironworking in the Middle Ages, benefits from a dense bibliography, which has already been discussed. We will confine ourselves here to reporting the latest findings on the application of hydraulic power to wind tunnels. From the 12th and 13th centuries onward, the iron mill, found in Burgundy, Champagne and Franche-Comté, consisted of a direct reduction forge equipped with a hydraulic hammer that struck the iron bloom coming out of the furnace to clean, refine and compact it. The mill, which became increasingly common in the Languedoc and Pyrenean valleys between the end of the 13th century and the beginning of the 14th century, was certainly equipped with a hydraulic hammer, as we have already mentioned. As for the hydraulic wind tunnel, which complements hammering, it has been proven in the forge of Léca (Pyrénées-Orientales) as early as 1416 (Verna 2017a, pp. 212–215). This application testifies to particular skills in the operation of the wind tunnel, as the melting point must not be reached, which presupposes the regularity and control of the blown air. At the same time, in his treatise *De rebus militaribus*, the Sienese engineer Mariano di Jacopo, known as Taccola, described a reduction forge with a hydraulic wind tunnel, a workshop he had no doubt observed. Furthermore, the first blast furnaces and refineries, documented in the kingdom of France as early as the end of the 14th century (the Walloon process), also used hydraulic wind tunnels. With the emergence of new hydraulic forges (bloomery or finery processes) that combine waterwheel-driven tools (hammer and bellows), an innovative threshold was crossed in the 1300s.

1.4. Mills and men

Influenced by the historical materialism and total history championed by Lucien Febvre and the *Annales* school, Marc Bloch's article is one of the founding texts of the social history of technology. In the 1960s, Charles Parain reinforced the Marxist approach by founding the history of mills on the analysis of productive forces and transformations in relations production (Parain 1965). In the 1980s, François Sigaut pursued this line of thought, focusing on the "actors" involved in milling, particularly the female workforce associated with the domestic sphere, which was replaced by men when the activity became mechanical and commercial (Sigaut 1986). At the same time, regional and monographic studies were multiplying, putting the relationship between mills and banal seigniory to the test. More recently, research into rural industry, particularly industrial mills, has explored the modalities of technical change in the countryside, highlighting the role of entrepreneurs, technicians and markets (Verna 2017a). For a generation now, historians have been looking at population growth and food consumption to understand the development and evolution of medieval mills (Comet 1992; Mouthon 1993; Arnoux 2012; Marchandin 2021); in fact, Charles Parain had already noted the need to consider the quality of cereals consumed and ground.

1.4.1. *From banal seigniory to medieval enterprise*

In 1935, Marc Bloch highlighted two periods in the adoption of hydraulic installations. The first, which he placed between late Antiquity and the High Middle Ages, was linked to the decline in slave labor (Bloch 1935a, p. 547); the second, between the 10th and 11th centuries, was linked to the establishment of the banal seigniory (Bloch 1935a, pp. 552–556). In 1965, Charles Parain extended the demonstration, emphasizing the alleged support of peasants, who, owing to the use of the lordship mill, benefited from time savings in the development of their farms (Parain 1965, pp. 66–67). In the early 1980s, Robert Fossier took up the issue again, both in its chronological aspect and in his analytical approach to feudal structures (Fossier 1982). On the basis of a corpus of written sources, he links the various waves of mill settlements to the "*encellulement des hommes*"[9] (Fossier 1982, p. 231), in other words, their supervision by the banal seigniory and the establishment of the village, between the 10th and 12th centuries. He established a link between the establishment of the feudal system and the movement to build mills

9 This concept is derived from that of *incastellamento*, developed by Pierre Toubert, to designate the grouping of people in fortified rural villages in Latium between the 10th and 12th centuries. It highlights the control of lords, which is not necessarily associated with a fortified settlement.

in the northern and northeastern regions (Picardy, Lorraine, Burgundy). In southern regions (Catalonia, Provence, Languedoc), he emphasizes that lords took advantage of existing hydraulic facilities to establish their authority. The mill was a tool for the exercise of power, as Pierre Bonnassie had shown a few years earlier (Bonnassie 1975). The originality of "Catalan feudalism" lies in the speed and violence with which it was established in the middle of the 11th century. The aristocracy sought to reserve the fruits of growth for itself. The mills belonging to the peasant farmers were destroyed and replaced by more powerful installations built by the lords. The authors mentioned above remind us that mills were part of the equipment, which, along with ovens, were part of a lordship monopoly (banality) imposed on peasants. Since then, studies have shown that banality is sometimes limited, or even non-existent, in both the northern and southern halves of France, as well as contrasting situations within a single region, such as in Quercy (Lartigaut 1978; Charbonnier 1980; Mouthon 1993; Larguier 1996).

Historians agree that building and maintaining a watermill is costly: the mechanisms are fragile, and the need for repairs is common, despite the recycling strategies implemented, including the reuse of runner millstones as nether millstones. In line with thinking on "mill equipment and manpower management", the question of financing hydraulic installations has long been limited to the following observation: watermills presuppose a social organization capable of building and maintaining them. Initially, this framework was limited to banal seigniory, as mentioned above. Since then, historians have shown that the canonical definition of banality does not consider the type of mill, the chronology of the construction and monopolization of hydraulic installations by the lords, or even less so the economic logics involved in defining, or redefining, the associated usage rights (Arnoux 2003, 2008, 2012). The Cistercian case offers a well-documented example of the dynamics at work in the choice of hydraulics. Since the 1980s, under the aegis of Paul Benoît, research into Cistercian abbeys has shed light on the role of the White Monks in the use and dissemination of hydraulic power in the Middle Ages. Studies initially focused on their mining and metallurgical activities (Verna 1995). The investigation has continued in the context of academic work (Rouzeau 2008, 2020) and collective research programs on monastic hydraulics in Burgundy, Champagne and Franche-Comté (Benoît and Berthier 1998; Berthier and Rouillard 1998) and, recently, on the landscapes, layout and architecture of Cistercian monasteries between the Seine and Rhine[10]. Today, it seems clear that the

10 "*Paysages et architecture des monastères cisterciens entre Seine et Rhin du* XII*e au* XVIII*e siècle*" (Landscapes and architecture of Cistercian monasteries between the Seine and Rhine from the 12th to 18th century) collective research program, coordinated by Agnès Charignon, Inrap.

Cistercians distinguished themselves by acquiring and building industrial mills as early as the 12th century, which should not be limited to hydraulic forges. At the same time, they owned fulling mills, which spread rapidly from Pontigny, and then, in the following centuries, bark and woad mills (Benoît and Berthier 1998; Berthier and Rouillard 1998; Baudin et al. 2019). Beyond Cîteaux's voluntarist policy, in the mechanization carried out by the order – which necessarily induces an increase in productivity – historians observe the abandonment of the initial objective of self-consumption to voluntarily open up and supply markets, both locally and internationally (Verna 1995; Arnoux 2012, pp. 332–333; Baudin et al. 2019). By moving forward in chronology, new configurations and new groups can be recorded. During the economic upturn of the 15th century, when the trend was for pre-existing mills to be brought back into service, ecclesiastical and secular lords implemented incentives by agreeing to tax arrangements attached to hydraulic equipment (Lartigaut 1978; Caliste 2021). At the time, investment was carried out by various actors who used their capital to acquire the means of production while assuming the costs of rebuilding and maintaining the mills.

Medievalists have often referred to watermills as "enterprises", with the most famous example being the mills of the Bazacle in Toulouse, illustrating complex financing (shareholding) and management (Sicard 1953). The analysis, initially devoted to the city and urban capital, shifted to the countryside from the late 1980s onward, since the proliferation of mills in the Middle Ages was clearly not an exclusively urban phenomenon, as regional studies have repeatedly demonstrated, like the Cistercian observatory just mentioned. While the movement led by the White Monks is the best known, it is not the only movement. The financial capacity of large abbeys of all orders was used to build and acquire mills or shares in mills, primarily but not exclusively for the production of wheat, as early as the mid-12th century (Benoît and Berthier 1998; Berthier and Rouillard 1998). In addition, the reflections of the last 10 years on enterprise and the entrepreneur in the medieval countryside offer a new reading grid (Minovez et al. 2013; Verna 2017a, 2017b). A workshop, which is a gristmill or industrial mill, can be considered an enterprise in the case where its production is open to the market and generates profit. The mouline is a well-documented example of this (Verna 2017a). In the southern part of the kingdom of France, and in the crown of Aragon, the establishment and rapid spread of the hydraulic forge was supported by royal and princely legislation, the reorganization of supply markets, particularly for combustibles, and a wartime economy. The notability of the small towns provided initial financing for the workshops (fixed capital), which were run by entrepreneurs, often foreigners (Basques, Italians). They would rent the forge under short-term leases. The entrepreneur would run the workshop, assemble and pay the workforce, manage access to raw materials and sell production on a wholesale market supervised by local notables.

1.4.2. *Men at work*

Medievalists have gradually moved closer to the people, closer to the installations. The microhistorical approach and the (re)discovery of sources, particularly practical acts, have enabled us to gain a better understanding of the individuals who designed hydraulic machines and, even more so, those who built and worked in them. It was not until the 15th century that technical treatises shed light on the configuration, and sometimes the design, of hydraulic machines – in this case, mills and their mechanisms – as described by early Renaissance engineers. Despite the known pitfalls – limited numbers, drawings and descriptions sometimes impossible to put into practice – it is a fact that some engineers set out to transpose what was observed (Comet 2001, 2003; Mane 2002). For example, Taccola presents five mills, some of which were civil use at the time of writing, in the first half of the 15th century. Another Sienese, Francesco di Giorgio Martini, dedicated a large part of his Treatise on Civil and Military Architecture, composed between 1470 and 1480, to civil mechanics, in particular mills, illustrated by over 50 drawings. Several German clerical engineers were also interested in hydraulic problems at the end of the Middle Ages (Lohrmann 2003, 2017). As Georges Comet, who took up the subject from the 1990s, noted, these documents reveal the expectations of the time, particularly the research carried out to increase the power of equipment, sometimes through the addition of energy sources or the use of new techniques, such as the connecting rod (Comet 2001, 2003). The works of the Germans Konrad Gruter de Werden and Strada de Rosberg, written in the 15th and 16th centuries, respectively, are thus marked by the technical challenges of their time, arising from the development of mining, with a particular focus on dewatering. Historians have long dismissed any link between the technical literature and the craftsmen who built watermills, owing to the supposed and exclusive target of technical treatises, that of the princely courts (Comet 2001 2003; Chandelier et al. 2017). Dietrich Lohrmann highlighted the relationships that the clerk engineer Gruter de Werden maintained with 10 mill builders (Lohrmann 2003, 2017). In addition, through the study of practice acts, medievalists have revealed other professionals whose expertise, function and status are the objects of social consensus. These individuals, appointed as experts to establish a valuation, or a financial estimate, are recognized for their knowledge and experience, which are acquired mostly through learning and imitation (Feller and Verna 2012). The individuals solicited to check and assess the condition of hydraulic equipment prior to the conclusion of a leasing contract are masons and carpenters, often specialized in mill construction (Marchandin 2021; Caliste 2021). They possess vernacular expertise, which is based on tacit knowledge, is not formulated in writing and is recognized by their contemporaries.

Among the workers who made use of hydraulic machines in the Middle Ages, millers – those who regulated and operated the mill's equipment to transform grain into flour – have received particular attention because of their social position. Operating equipment associated with seigniorial duties thus became the emblematic figure of banal seigniory (Rivals 2000). In recent years, medievalists have reconstructed biographies to assess the activities, relationships and wealth of an individual or group (Bernardi 2000; Verna 2017a). These works show a reality somewhat close to that reported in the medieval literature (Ménard 2002) – some millers were wealthy enough to rent several mills – but also more complex. Regardless of whether they operated the mills themselves, the millers had knowledge of hydraulics and milling. The question of how they acquired their technical knowledge remains to be clarified, given the limited number, or even absence, of apprenticeship contracts. Some millers declared themselves to be mill builders. They selected the millstones, refurbished them and ensured that they were properly spaced. They were responsible for preventing breakdowns and carrying out the most common repairs. The economic and social ascent of miller families in Paris at the end of the Middle Ages has been documented, with this trend accelerating during the 16th century (Marchandin 2021).

As far as industrial mills are concerned, historians have been able to identify a number of individuals, highlighting the movement between technical territories. At the end of the Middle Ages, this was the case for the moulines in Vallespir (Catalonia). The workers there in the 15th century were Basques, whose skills were sought after by the local lessors, who rented out the workshops to them. This workforce, plagued by chronic indebtedness, faced competition in the 1500s from Italians from the north of the peninsula (Verna 2017a, 2023). The latter benefitted from a favorable labor market, linked to the rebuilding of forges and the evolution of the ironworking process with the introduction of trip hammers dedicated to the production of iron wire. Similarly, the biographical approach has enabled us to highlight economic strategies in Languedoc, in the field of woollen cloth and its mechanical fulling (Caliste 2021). Involved in the workshop – the fulling mill often associated with a dyeing plant – Languedoc fullers were major players in the marketing of the region's woollen cloth. These entrepreneurs took advantage of their technical skills and networks to reach regional and international markets. While it is possible to follow the careers and activities of these men, medievalists are unable to record their daily lives in the workshop due to a lack of texts or images. Historians have therefore turned to ethnology and experimental archaeology to try and retrace technical actions.

Thus, since 1935, when Marc Bloch's article was published, "the reading of which is never finished" (Lohrmann 1990), research into watermills has undergone considerable renewal. The temporality of innovations linked to the use of hydraulic power has been disrupted. The common use of watermills from the 1st century CE is no longer a matter of debate. The widespread use of gristmills during the High Middle Ages is also a proven fact, both in written sources and in archaeological excavations. However, more needs to be said about industrial mills. In particular, while the material aspects of gristmills – hydraulic installations, engines and tools – are now better understood owing to the combined efforts of textual historians and archaeologists, there are still large gray areas concerning the technical aspects of industrial mills, with the exception of hydraulic forges. The close link between societies and the techniques mentioned by Marc Bloch has also been explored by several generations of medievalists. The mill found its place in banal seigniory and in villages. More recently, interest in markets and in the diversity and quality of the raw materials processed has once again brought historians back to the hydraulic machine and its mechanisms. Finally, the question of financing mills – fragile and therefore costly installations – has enabled us to go beyond the mythical figure of the miller to approach the craftsman, the technician and the investor. Historians have taken a deep dive into the workshops, putting the man at work at the heart of the history of technology, as Marc Bloch and Lucien Febvre hoped.

1.5. References

Amouric, H. (1980). Innovation et invention, un exemple provençal XVe-XVIIIe siècles. *Technologies, idéologies, pratiques*, 2(1), 1–11.

Amouric, H. (1984). Moulins et meunerie en Basse-Provence occidentale du Moyen Âge à l'ère industrielle. PhD Thesis, Université de Provence, Aix-en-Provence.

Arnoux, M. (1993). *Mineurs, férons et maîtres de forge : études sur la production du fer dans la Normandie du Moyen Âge, XIe-XVe siècles*. Éditions du CTHS, Paris.

Arnoux, M. (1994). Moulins à fer et procédé indirect. Énergie hydraulique et innovation dans la sidérurgie européenne (XIIe-XVIe siècles). In *Le contrôle des eaux en Europe occidentale XIIe-XVIe siècle. Proceedings Eleventh International Economic History Congress*, Crouzet-Pavant, E. and Maire-Vigueur, J.C. (eds). Bocconi University, Milan.

Arnoux, M. (2003). Moulins seigneuriaux et moulins royaux en Normandie : marché de l'énergie et institutions (XIIe-XVe siècles). In *Economia e energia, secc. XIII-XVIII. Atti della trentaquattresima settimana di studio dell'Istituto internazionale di storia economica "F. Datini" di Prato, 15-19 April 2002*. Le Monnier, Florence.

Arnoux, M. (2008). Les moulins à eau en Europe occidentale (IXe-XIIe siècles). Aux origines d'une économie institutionnelle de l'énergie hydraulique. In *L'acqua nei secoli altomedievali. Atti della LV Settimana di studio (Spoleto, 12-17 April 2007)*. Fondazione Centro italiano di studi sull'alto Medioevo, Spoleto.

Arnoux, M. (2012). *Le temps des laboureurs : travail, ordre social et croissance en Europe (XIe-XIVe siècle)*. Albin Michel, Paris.

Arribet-Deroin, D. (1998). Un haut fourneau primitif en Pays de Bray : Glinet (Seine-Maritime). In *L'Innovation technique au Moyen Âge. Actes du VIe congrès international d'archéologie médiévale*, Beck, P. (ed.). Éd. Errance, Paris.

Bailly-Maître, M.-C. (ed.) (2019). *L'entreprise minière de Brandes XIe-XIVe siècles. Huez, Oisans (Isère). Extraction et transformation de minerais argentifères*. Alpara, Lyon.

Baudin, A., Benoît, B., Rouillard, J., Rouzeau, B. (eds) (2019). *L'industrie cistercienne XIIe-XXIe siècles*. Somogy Éditions d'art, Paris.

Bautier, A.M. (1960). Les plus anciennes mentions de moulins hydrauliques industriels et de moulins à vent. *Bulletin philologique et historique (jusqu'à 1610) du comité des travaux historiques et scientifiques*, II, 567–626.

Beck, P. (ed.) (1998). *L'innovation technique au Moyen Âge. Actes du VIe congrès international d'archéologie médiévale*. Éd. Errance, Paris.

Belhoste, J.F. (1992). *Fonte, fer, acier : Rhône-Alpes (XVe-début XXe siècle)*. Association pour le développement de l'Inventaire des richesses artistiques de Rhône-Alpes, Lyon.

Belhoste, J.F., Lecherbonnier, Y., Arnoux, M. (1991). *La métallurgie normande (XIIe-XVIIe siècles) : la révolution du haut fourneau*. Association Histoire et Patrimoine Industriels de Basse-Normandie, Caen.

Belhoste, J.F., Claerr-Roussel, C., Lassus, F., Corbineau, F., Priou, J.F., Brugier, M. (1994). *La métallurgie comtoise (XVe-XIXe siècles) : étude du Val de Saône*. ASPRODIC, Besançon.

Belmont, A., Jaccottey, L., Lepareux-Couturier, S. (2016). Les meules à grain de l'époque médiévale et de la Renaissance : première approche. In *Archéologie des moulins hydrauliques, à traction animale et à vent des origines à l'époque médiévale et moderne en Europe et dans le monde méditerranéen. Actes du colloque international, Lons-le-Saunier du 2 au 5 novembre 2011*, Jaccottey, L. and Rollier, G. (eds). Presses Universitaires de Franche-Comté, Besançon.

Benoît, P. (1997). *La mine de Pampailly, XVe-XVIIIe siècles (Brussieu-Rhône)*. Alpara, Lyon.

Benoît, P. (1998). Au four et au moulin : innovation et conjoncture. In *L'innovation technique au Moyen Âge. Actes du VIe congrès international d'archéologie médiévale*, Beck, P. (ed.). Éd. Errance, Paris.

Benoît, P. (2016). Remarques sur les fouilles de moulins à eau médiévaux en Europe. In *Archéologie des moulins hydrauliques, à traction animale et à vent des origines à l'époque médiévale et moderne en Europe et dans le monde méditerranéen. Actes du colloque international, Lons-le-Saunier du 2 au 5 novembre 2011*, Jaccottey, L. and Rollier, G. (eds). Presses Universitaires de Franche-Comté, Besançon.

Benoît, S. (2020). *D'eau et de feu : forges et énergie hydraulique (XVIIIe-XXe siècle). Une histoire singulière de l'industrie française*. PUR, Rennes.

Benoît, P. and Berthier, K. (1998). L'innovation dans l'exploitation de l'énergie hydraulique d'après le cas des monastères cisterciens de Bourgogne, Champagne et Franche-Comté. In *L'innovation technique au Moyen Âge. Actes du VIe congrès international d'archéologie médiévale*, Beck, P. (ed.). Éd. Errance, Paris.

Benoît, P. and Cailleaux, D. (eds) (1991). *Moines et métallurgie dans la France médiévale*. Association pour l'édition et la diffusion des études historiques, Paris.

Bernardi, P. (2000). Essai, tâtonnement et pari : le rôle de l'individu dans l'innovation. *Médiévales, Techniques : les paris de l'innovation*, 39, 14–29.

Bernardi, P. and Verna, C. (2001). Travail et Moyen Âge : un renouveau historiographique. *Cahiers d'histoire, revue d'histoire critique*, 83, 27–46.

Berthier, K. and Rouillard, J. (1998). Nouvelles recherches sur l'hydraulique cistercienne en Bourgogne, Champagne et Franche-Comté. *Archéologie médiévale*, 28, 121–147.

Bloch, M. (1935a). Avènement et conquêtes du moulin à eau. *Annales d'histoire économique et sociale*, 7, 538–563.

Bloch, M. (1935b). Les "inventions" médiévales. *Annales d'histoire économique et sociale*, 7, 634–644.

Bolòs, J. (2002). Les moulins en Catalogne au Moyen Âge. In *Moulins et meuniers dans les campagnes européennes : IXe-XVIIIe siècle. Actes des XXIe Journées internationales d'histoire de l'abbaye de Flaran, septembre 1999*, Mousnier, M. (ed.). Presses universitaires du Mirail, Toulouse.

Bonnassie, P. (1975). *La Catalogne du milieu du Xe à la fin du XIe siècle : croissance et mutations d'une société*. Association des publications de l'Université de Toulouse-Le Mirail, Toulouse.

Braunstein, P. (1998). Savoir et savoir-faire : les transferts techniques. In *L'innovation technique au Moyen Âge. Actes du VIe congrès international d'archéologie médiévale*, Beck, P. (ed.). Éd. Errance, Paris.

Brun, J.P. (2016). Les moulins hydrauliques dans l'Antiquité. In *Archéologie des moulins hydrauliques, à traction animale et à vent des origines à l'époque médiévale et moderne en Europe et dans le monde méditerranéen. Actes du colloque international, Lons-le-Saunier du 2 au 5 novembre 2011*, Jaccottey, L. and Rollier, G. (eds). Presses Universitaires de Franche-Comté, Besançon.

Buchsenschutz, O., Chaigneau, C., Lepareux-Couturier, S. (2020). L'évolution des moulins dans la problématique de la production, de la commercialisation et de la consommation. In *Le changement dans les économies antiques*, Lerouxel, F. and Zurbach, J. (eds). Ausonius, Bordeaux.

Caliste, L. (2021). Le Lodévois et ses marges. Industries, ateliers et entreprises (vallées de la Lergue et de l'Hérault, XVe siècle). PhD Thesis, Université Paris 8, Vincennes – Saint-Denis.

Cardon, D. (1996). Le moulin à foulon : double fonction, double source de gain en productivité. *Arts et Traditions Rurales*, 16, 107–118.

Carus-Wilson, E. (1941). An industrial revolution of the 13th century. *Economic Historical Review*, 11, 39–60.

Caucanas, S. (1995). *Moulins et irrigation en Roussillon du IXe au XVe siècle*. CNRS Éditions, Paris.

Champion, E. (1996). *Moulins et meuniers carolingiens dans les polyptyques entre Loire et Rhin*. Association pour l'edition et la diffusion des études historiques-Vulcain, Paris.

Chandelier, J., Verna, C., Weill-Parot, N. (eds) (2017). *Science et technique au Moyen Âge (XIIe-XVe siècles)*. Presses Universitaires de Vincennes, Saint-Denis.

Charbonnier, P. (1980). *Une autre France : la seigneurie rurale en basse Auvergne du XIVe au XVIe siècle*. Institut d'études du Massif central, Clermont-Ferrand.

Comet, G. (1992). *Le paysan et son outil : essai d'histoire technique des céréales, France, VIIIe-XVe siècle*. École française de Rome, Rome.

Comet, G. (2001). Des machines pour quoi faire ? Quelques réflexions à propos des ingénieurs à la Renaissance. In *Techniques et sociétés en Méditerranée. Hommage à Marie-Claire Amouretti*, Brun, J.P. and Jockey, P. (eds). Maisonneuve et Larose, Paris.

Comet, G. (2002). Moulins et meuniers : réflexions historiographiques et méthodologiques. In *Moulins et meuniers dans les campagnes européennes : IXe-XVIIIe siècle. Actes des XXIe Journées internationales d'histoire de l'abbaye de Flaran, septembre 1999*, Mousnier, M. (ed.). Presses universitaires du Mirail, Toulouse.

Comet, G. (2003). Pour une histoire des moulins entre technique et idéologie In *I mulini nell'Europa medievale: Atti del convegno di San Quirino d'Orcia, 21-23 September 2000*, Galetti, P. and Racine, P. (eds). CLUEB, Bologna.

Daumas, M. (ed.) (1962). *Histoire générale des techniques. Les origines de la civilisation technique, t. I*. PUF, Paris.

Daumas, M. (ed.) (1965). *Histoire générale des techniques. Les premières étapes du machinisme, t. II*. PUF, Paris.

Dillmann, P. and L'Héritier, M. (2016). Archéométrie et histoire des techniques : les procédés direct et indirect en sidérurgie (XIVe-XVIIe siècle). *Artefact*, 4, 63–81.

Durand, A. (2002). Les moulins carolingiens du Languedoc (fin VIIIe-début XIe siècle). In *Moulins et meuniers dans les campagnes européennes : IXe-XVIIIe siècle. Actes des XXIe Journées internationales d'histoire de l'abbaye de Flaran, septembre 1999*, Mousnier, M. (ed.). Presses universitaires du Mirail, Toulouse.

Febvre, L. (1935). Réflexions sur l'histoire des techniques. *Annales d'histoire économique et sociale*, 7, 531–535.

Feller, L. and Verna, C. (2012). Expertises et cultures pratiques. In *Experts et expertises au Moyen Âge. Consilium quaeritur a perito, XLIIe congrès de la SHMESP, Oxford, 2011*. Publications de la Sorbonne, Paris.

Fossier, R. (1982). L'équipement en moulins et l'encadrement des hommes. In *L'histoire des sciences et des techniques doit-elle intéresser les historiens ? Colloque organisé par la Société française d'Histoire des Sciences et des Techniques*. Centre de recherches historiques et juridiques de l'Université Paris, Paris.

Gille, B. (1954). Le moulin à eau, une révolution technique médiévale. *Techniques et civilisations*, III, 1–15.

Gille, B. (1960). Les origines du moulin à fer. *Revue d'histoire de la sidérurgie*, I–3, 24–32.

Gille, B. (ed.) (1978). *Histoire des techniques : technique et civilisations, technique et sciences*. Éditions Gallimard, Paris.

Hilaire-Pérez, L. and Verna, C. (2016). Histoire économique et histoire des techniques (XVe-XVIIIe siècle). *Artefact*, 4, 13–30.

Jaccottey, L. and Rollier, G. (eds) (2016a). Archéologie des moulins hydrauliques, à traction animale et à vent des origines à l'époque médiévale et moderne en Europe et dans le monde méditerranéen. In *Archéologie des moulins hydrauliques, à traction animale et à vent des origines à l'époque médiévale et moderne en Europe et dans le monde méditerranéen. Actes du colloque international, Lons-le-Saunier du 2 au 5 novembre 2011, volume 2*. Presses Universitaires de Franche-Comté, Besançon.

Jaccottey, L. and Rollier, G. (eds) (2016b). Archéologie des moulins hydrauliques, à traction animale et à vent des origines à l'époque médiévale et moderne en Europe et dans le monde méditerranéen. Bilan et perspectives. In *Archéologie des moulins hydrauliques, à traction animale et à vent des origines à l'époque médiévale et moderne en Europe et dans le monde méditerranéen. Actes du colloque international, Lons-le-Saunier du 2 au 5 novembre 2011*, volume 2. Presses Universitaires de Franche-Comté, Besançon.

Jarrige, F. (2023). *La ronde des bêtes : le moteur animal et la fabrique de la modernité*. La Découverte, Paris.

Larguier, G. (1996). *Le drap et le grain en Languedoc : Narbonne et Narbonnais 1300-1789*. Presses Universitaires de Perpignan, Perpignan.

Lartigaut, J. (1978). *Les campagnes du Quercy après la Guerre de Cent ans (vers 1440-vers 1500)*. Association des publications de l'Université de Toulouse-le-Mirail, Toulouse.

Le Goff, J. (1964). *La civilisation de l'Occident médiéval*. Arthaud, Paris.

Lohrmann, D. (1990). L'histoire du moulin à eau, avant et après Marc Bloch. In *Marc Bloch aujourd'hui. Histoire comparée et sciences sociales*, Hartmut, A. and Burguière, A. (eds). Éditions de l'École des Hautes Études en Sciences Sociales, Paris.

Lohrmann, D. (2003). Les moulins d'un ingénieur allemand en Italie, vers 1425. In *I mulini nell'Europa medievale: Atti del convegno di San Quirino d'Orcia, 21-23 septembre 2000*, Galetti, P. and Racine, P. (eds). CLUEB, Bologna.

Lohrmann, D. (2017). Théorie et pratique de l'hydraulique chez Konrad Gruter, clerc et technicien allemand en Italie (1393-1424). In *Science et technique au Moyen Âge (XIIe-XVe siècles)*, Chandelier, J., Verna, C., Weill-Parot, N. (eds). Presses Universitaires de Vincennes, Saint-Denis.

Longepierre, S. (2012). *Meules, moulins et meulières en Gaule méridionale du IIe s. av. J-C. au VIIe s. ap. J-C.* Mergoil, Montagnac.

Lucas, A. (2016). Fulling mills in Medieval Europe: Comparing the manuscript and archaeological evidence. In *Archéologie des moulins hydrauliques, à traction animale et à vent des origines à l'époque médiévale et moderne en Europe et dans le monde méditerranéen. Actes du colloque international, Lons-le-Saunier du 2 au 5 novembre 2011*, volume 2, Jaccottey, L. and Rollier, G. (eds). Presses Universitaires de Franche-Comté, Besançon.

Mane, P. (2002). Les moulins à eau dans l'iconographie médiévale. In *Moulins et meuniers dans les campagnes européennes : IXe-XVIIIe siècle. Actes des XXIe Journées internationales d'histoire de l'abbaye de Flaran, septembre 1999*, Mousnier, M. (ed.). Presses universitaires du Mirail, Toulouse.

Marchandin, P. (2021). Moulins et énergie à Paris du XIIIe au XVIe siècle. PhD Thesis, Université Paris Sciences et Lettres, École nationale des chartes, Paris.

Ménard, P. (2002). Moulins et meuniers dans la littérature médiévale. In *Moulins et meuniers dans les campagnes européennes : IXe-XVIIIe siècle. Actes des XXIe Journées internationales d'histoire de l'abbaye de Flaran, septembre 1999*, Mousnier, M. (ed.). Presses universitaires du Mirail, Toulouse.

Mille, P. (2016). Évolution technique des moulins à roues verticales en dessous, mis au jour sur le territoire français, VIIIe-XIVe siècles. In *Archéologie des moulins hydrauliques, à traction animale et à vent des origines à l'époque médiévale et moderne en Europe et dans le monde méditerranéen. Actes du colloque international, Lons-le-Saunier du 2 au 5 novembre 2011*, volume 2, Jaccottey, L. and Rollier, G. (eds). Presses Universitaires de Franche-Comté, Besançon.

Minovez, J.M., Verna, C., Hilaire-Pérez, L. (2013). L'industrie des campagnes : retrouver la longue durée. In *Les Industries rurales dans l'Europe médiévale et moderne. Actes des XXXIIIe journées internationales d'histoire de l'abbaye de Flaran, 7-8 octobre 2011*, Minovez, J.M., Verna, C., Hilaire-Pérez, L. (eds). Presses universitaires du Mirail, Toulouse.

Minvielle-Larousse, N., Jaccottey, L., Lepareux-Couturier, S. (2017). Les outils de mouture en Champagne-Ardenne à l'époque médiévale. In *Les meules du Néolithique à l'époque médiévale : technique, culture, diffusion. Actes du 2e colloque du Groupe Meule, Reims, du 15 au 17 mai 2014*, Buchsenschutz, O., Lepareux-Couturier, S., Fronteau, G. (eds). S.A.E., Dijon.

Moritz, L.A. (1958). *Grain Mills and Flour in Classical Antiquity*. Clarendon Press, Oxford.

Mousnier, M. (ed.) (2002). *Moulins et meuniers dans les campagnes européennes : IXe-XVIIIe siècle. Actes des XXIe Journées internationales d'histoire de l'abbaye de Flaran, septembre 1999*. Presses universitaires du Mirail, Toulouse.

Mouthon, F. (1993). Les blés bordelais : l'économie céréalière dans les diocèses de Bordeaux et de Bazas (vers 1300-vers 1550). PhD Thesis, Université Bordeaux 3, Bordeaux.

Pagès, G. and Verna, C. (2022). L'invention de l'industrie antique et médiévale. *Artefact*, 17, 243–264.

Parain, C. (1965). Rapports de production et développement des forces productives : l'exemple du moulin à eau. *La pensée*, 119, 55–70.

Phalip, B. (1992). Le moulin à eau médiéval. Problème et apport de la documentation languedocienne. Bassins de l'Hérault, Orb et Vidourle. *Archéologie du Midi Médiéval*, 10, 63–96.

Rivals, C. (1984). Divisions géographiques de la France indiquées par une analyse des moulins en 1809. *Revue géographique des Pyrénées et du Sud-Ouest*, 55(3), 367–384.

Rivals, C. (2000). *Le Moulin et le meunier. Mille ans de meunerie en France et en Europe. Technique, Société, Culture, Patrimoine*. Éd. Empreinte, Portet-sur-Garonne.

Roch, J.-L. (2013). *Un autre monde du travail : la draperie en Normandie au Moyen Âge*. Presses universitaires de Rouen et du Havre, Mont-Saint-Aignan.

Rohmer, P., Jaccottey, L., Mille, P., Gluhak, M.T. (2016). Les moulins carolingiens (IXe-Xe siècles) d'Audun-le-Tiche (Moselle). In *Archéologie des moulins hydrauliques, à traction animale et à vent des origines à l'époque médiévale et moderne en Europe et dans le monde méditerranéen. Actes du colloque international, Lons-le-Saunier du 2 au 5 novembre 2011*, volume 1, Jaccottey, L. and Rollier, G. (eds). Presses Universitaires de Franche-Comté, Besançon.

Rollier, G. (1998). Aménagements hydrauliques : le cas de Cluny. In *L'Innovation technique au Moyen Âge. Actes du VIe congrès international d'archéologie médiévale*, Beck, P. (ed.). Éd. Errance, Paris.

Rollier, G., Jaccottey, L., Mille, P., Girarclos, O. (2016). Les fouilles du moulin de Thervay : évolution d'un site de meunerie de la période carolingienne à l'installation du domaine de l'abbaye cistercienne d'Acey (IXe-XIIe siècles). In *Archéologie des moulins hydrauliques, à traction animale et à vent des origines à l'époque médiévale et moderne en Europe et dans le monde méditerranéen. Actes du colloque international, Lons-le-Saunier du 2 au 5 novembre 2011*, volume 1, Jaccottey, L. and Rollier, G. (eds). Presses Universitaires de Franche-Comté, Besançon.

Rouillard, J. (2011). De l'acier pour les moulins (Bourgogne et Champagne, XIVe-XVe siècle). In *L'Acier en Europe avant Bessemer*, Dillmann, P., Hilaire-Pérez, L., Verna, C. (eds). Presses universitaires du Midi, Toulouse.

Rouillard, J. (2016). L'apport des sources écrites à l'archéologie du moulin à eau médiéval. In *Archéologie des moulins hydrauliques, à traction animale et à vent des origines à l'époque médiévale et moderne en Europe et dans le monde méditerranéen. Actes du colloque international, Lons-le-Saunier du 2 au 5 novembre 2011*, volume 2, Jaccottey, L. and Rollier, G. (eds). Presses Universitaires de Franche-Comté, Besançon.

Rouzeau, B. (2008). Le patrimoine hydraulique et industriel de l'abbaye cistercienne de Morimond entre la fondation et la Guerre de Trente Ans : énergies et matériaux d'après les sources écrites et archéologiques. PhD Thesis, Université Paris 1 Panthéon-Sorbonne, Paris.

Rouzeau, B. (2020). Paul Benoît et les Cisterciens : des techniques à l'économie, une page d'histoire. Bibliographie de Paul Benoît. In *Les cisterciens et l'eau. Hommage à Paul Benoît*, Rouzeau, B. and Grelois, A. (eds). Commentarii cisterciences, Cîteaux.

Schumpeter, J.A. (1935). Théorie de l'évolution économique. In *Recherches sur le profit, le crédit, l'intérêt et le cycle de la conjoncture*, Anstett, J.J. (trans.). Dalloz, Paris.

Sicard, G. (1953). *Les moulins de Toulouse au Moyen Âge : aux origines des sociétés anonymes*. Armand Colin, Paris.

Sigaut, F. (1986). Moulins, industrie et sociéte. *Culture technique*, 16, 215–223.

Spain, R. (2008). *The Power and Performance of Roman Water-mills. Hydromechanical Analysis of Vertical-wheeled Water-mills*. BAR, Oxford.

Téreygeol, F. (forthcoming). "Bref ce sont les Indes Françoises..." (Malus, 1601) : mines et ateliers métallurgiques à Castel-Minier (X^e^-XVIe siècles).

Verna, C. (1995). *Les mines et les forges des Cisterciens en Champagne méridionale et en Bourgogne du Nord, XIIe-XVe siècle*. AEDEH, Paris.

Verna, C. (2000). Réduction du fer et innovation : à propos de quelques débats en histoire sociale des techniques. *Médiévales, Techniques : les paris de l'innovation*, 39, 79–95.

Verna, C. (2001). *Le temps des moulines. Fer, technique et société dans les Pyrénées centrales (XIIIe-XIVe siècles)*. Éditions de la Sorbonne, Paris.

Verna, C. (2008). "Moulin à fer". L'héritage de Bertrand Gille. In *Jeux d'eau : moulins, meuniers et machines hydrauliques, XIe-XXe siècle. Études offertes à Georges Comet*, Durand, A. (ed.). Publications de l'Université de Provence, Aix-en-Provence.

Verna, C. (2013). Pour une approche biographique de l'entreprise rurale au Moyen Âge. In *Les industries rurales dans l'Europe médiévale et moderne. Actes des XXXIIIe journées internationales d'histoire de l'abbaye de Flaran, 7-8 octobre 2011*, Minovez, J.M., Verna, C., Hilaire-Pérez, L. (eds). Presses universitaires du Mirail, Toulouse.

Verna, C. (2017a). *L'industrie au village. Essai de micro-histoire (Arles-sur-Tech, XIVe et XVe siècles)*. Les Belles Lettres, Paris.

Verna, C. (2017b). Entreprises rurales en Méditerranée occidentale (XIIIe-XVe siècles). Un programme collectif de recherche. In *Industrias y mercados rurales en los reinos hispánicos (siglos XIII-XV)*, volume 9, Navarro Espinach, G. and Villanueva Morte, C. (eds). Monografías de la Sociedad Española de Estudios Medievales, Murcia.

Verna, C. (2023). Travail et capital : la migration des Italiens dans les forges de Catalogne (vers 1489 – vers 1550). In *Les fruits de la terre. Études d'histoire médiévale offertes à Laurent Feller*, Dejoux, M., Dewez, H., Huertas, É. (eds). Éditions de la Sorbonne, Paris.

Verna, C. and Dillmann, P. (2018). Histoire des techniques médiévales et culture matérielle. In *La culture matérielle : un objet en question. Anthropologie, archéologie et histoire. Actes du colloque international de Caen (9 et 10 octobre 2015)*, Bourgeois, L., Alexandre-Bidon, D., Feller, L., Perrine, M., Verna, C., Wilmart, M. (eds). Presses universitaires de Caen, Caen.

Wikander, Ö. (ed.) (2000). *Handbook of Ancient Water Technology*. Brill, Boston.

Wilson, A. (2020). Roman water-power. Chronological trends and geographical spread. In *Capital, Investment, and Innovation in the Roman World*, Erdkamp, P., Verboven, K., Zuiderhoek, A. (eds). Oxford University Press, Oxford.

Woronoff, D. (1998). Le quotidien des techniques : de la répétition aux aménagements. In *Les entreprises et leurs réseaux : hommes, capitaux techniques et pouvoirs XIXe-XXe siècles*, Merger, M. and Barjot, D. (eds). Presses de l'Université de Paris-Sorbonne, Paris.

2

China and the Global History of Techniques

Delphine SPICQ
Collège de France-Université PSL, EHESS, Université Paris Cité, Paris, France

2.1. Introduction

The history of techniques is a relatively recent discipline. It was initially established with reference to Western history. Over the past 20 years, however, Asia and China, in particular, have played increasingly significant roles in the history of techniques.

The history of techniques in China has undergone an evolution similar to that of the history of techniques in other regions of the world, particularly in the West, namely from a diffusionist, Europe-centric model of universal history to one that approaches techniques from the perspective of its circulation, recently giving way to a global vision of techniques that emphasizes particular themes as explanatory factors of technical evolutions, whether convergent or divergent, between different regions or countries of the world. Each of these paradigm shifts requires us to question the previous model and modify the framework of thought, methodology and angle of reflection, as well as to broaden the range of sources (documentary or otherwise).

We propose a chronological presentation of the development of the history of techniques in China, in line with the evolution of the questions presented by the discipline we have just briefly outlined. In particular, we aim to show the contribution of reflections on the history of this part of the world to the discipline as a whole.

2.2. Diffusionist model

The diffusionist model, which prevailed until the mid-1990s, gives a central place to technical innovation. Technical innovation is the measure of the advancement of civilizations, and its study mainly consists of listing inventions and their authors. In this model of edifying and exclusively male-dominated history, the "classic" history of techniques is punctuated by great discoveries made by great men. Great murals of progress are painted, measured against great inventors and inventions. The performance of countries is assessed in terms of economic development based on the transformation and appropriation of nature through technical innovations.

This model, which aims to apply to the entire planet, proposes a universal history in which the spread of knowledge (and techniques) takes place from one cultural area to another according to stages of development based on Rostow's model. Disseminated in the 1960s through his book *The Stages of Economic Growth: A Non-Communist Manifesto* (a highly militant work aimed at warding off the spread of communism around the world and proposing a five-stage model of economic development), Walt Whitman Rostow's work focused on the transfer of technical innovations to less economically advanced countries, owing to their technological gap, to enable them to develop economically according to the European model. In this way, technical transfers ensure the conditions for economic take-off. This diffusionist model establishes the dominance of the West and Europe in particular (we study the great inventions of Europe and of the English Industrial Revolution in particular), but it does not explain other technological trajectories in the rest of the world. It also focuses exclusively on the economic side of techniques, only proposing economic explanations for the European exception and the role of technical progress in it.

In the 1960s, Joseph Needham, a historian of China, began to challenge this model. Originally, a renowned embryologist and biochemist at Cambridge University, Joseph Needham became fascinated by China and the history of its science and techniques after a long stay in the Middle Kingdom during the Second World War as a research scientist.

Needham retained the universalist aspect of the diffusionist model but shifted the focus to Asia and questioned Western superiority in science and technology. He asserts the universality of modern science, to which all the countries of the world have contributed through their traditional sciences. As a fervent advocate of rehabilitating Asia's place, and China in particular, in the history of science and technology, he never ceases to demonstrate – notably, with the writing of what has become the reference collection on the history of science and technology in China, the monumental work *Science and Civilisation in China* – that the world and the West owe a great deal to the Chinese: the invention of paper, the compass, gunpowder and so on.

Although he makes room for other cultures, Needham maintains a classical teleological vision, proposing a linear trajectory of progress leading to the industrial revolution, with a frame of reference based on the history of European science and techniques and a sense of history that was to bring modern science and techniques to Europe. His famous question, the "Needham Puzzle", testifies to this: he wonders why China, which was ahead of Europe until the 15th century in terms of science and techniques, did not experience a scientific and then industrial revolution comparable to the one in Europe.

He himself attempted to answer this question and attributed the delay to the organization of Chinese bureaucratic society, which did not allow for the birth of an independent, dynamic merchant class that would have fostered economic expansion and the pursuit of technical inventions to develop trade and industry. The Chinese literate elites, who passed the civil service examinations and then administered the empire for the dynasty, were educated in strict Confucian orthodoxy and were careful to reproduce a moral order that was based on an agrarian society, guided by a traditional culture that left little room for innovation and shunned merchants.

Other historians of China, such as Roy Bin Wong and Kenneth Pomeranz of the California School, have also attempted to answer the Needham Puzzle. Pomeranz, in his 2000 book *The Great Divergence: China, Europe, and the Making of the Modern World Economy*, starts from the idea that until the middle of the 18th century, there was virtually no difference between the economies of the central regions of Europe (Great Britain and the Netherlands) and the advanced regions of China, such as the Yangtze Delta (the Lower Yangtze), which were comparable in size. He explains the British, and later European advancement, from the mid-18th century onward, was the result of access to coal and then raw materials from the colonies. This thesis is based on the assumption that England and the Lower Yangtze were at similar levels of economic and commercial development, a claim that has been widely criticized by many economic historians, including Jan de Vries, who believes that Pomeranz's arguments and data are unconvincing and that the Lower Yangtze and Europe were in no way economically comparable at the beginning of the 19th century: social and economic structures and levels of development differed enormously.

Wong also attempts to demonstrate that there is a comparable level of development between the two regions (Wong 1997). He analyzes the similarities and then the differences between Europe and Jiangnan and, like Pomeranz, demonstrates that these two regions are quite similar in terms of production (agricultural, artisanal or commercial), credit organization or per capita income, and social structures, but in terms of political structures, they differ completely. He explains the industrial take-off of the UK and Europe by a combination of technical innovation and the opening up of new energy sources. However, this does not explain where these technical innovations came from, or rather why they occurred in Europe and not in China.

In a second work, following de Vries' criticisms, Wong insists that differences in wage levels (higher in Europe than in China because the former is more urbanized and has a higher concentration of commerce and industry) explain technological innovations from the mid-18th century in Europe (Rosenthal and Wong 2011). Conversely, in China, low wages do not encourage the replacement of human labor by machines, nor do they lead to technical innovation.

Furthermore, Nathan Sivin, a close collaborator of Needham, questioned the latter's approach, asserting that the question of why the Industrial Revolution took place in Europe and not in China makes no sense, since it presupposes the advent of modern Western science (Sivin 1982). Beyond the teleological aspect, he also questions the conceptual framework that implicitly presents the superiority of Western values and traditions as the benchmark for any comparison and the universality of the terms used to compare categories that do not necessarily overlap. Indeed, the terms used, such as "science" or "techniques", are geographically and historically dated and cannot be transposed in space or time. The terms "science" or "techniques" exist in modern Chinese and cover the same meanings as their English or French equivalents but did not exist in Chinese before the last quarter of the 19th century. Does this mean that before the end of the 19th century, Chinese people did not know or practice scientific or technical activities and did not question the latter?

More generally, this linear, universalist model of technical diffusion has been criticized, above all, from the technical transfer perspective: many authors insist that the latter is never a simple replication, and its explanatory capacity is therefore weak. Several authors, including Nathan Rosenberg, explain that technical transmissions are uncertain and complex processes (Rosenberg 1970). Techniques are linked not only to technical progress but also to the places where they are received, the people who transmit them and so on. All of these criticisms of the diffusionist model and the reflections they provoked contributed to a rethinking of the analytical framework and the emergence of a new model in the early 2000s, that of circulation.

2.3. Circulation model (of techniques) and the cultural turn

This new explanatory model for the evolution of techniques and their dissemination is based on a critique of Pomeranz's concept of divergence, which is still based on the divide between industrial and non-industrial development, with a positive value given to the former. The advantage of this concept of divergence[1] was that it reintroduced techniques into history in a way that was both more local and more global and it allowed us to think more about the economy than about techniques.

This model, which rejects the concept of transfer and which is too closely linked to the logics of Western economic power, breaks with the global vision of the spread of knowledge from one cultural area to another: it starts locally and follows the circulation of a product or technique. It thus emphasizes the mobility of people, techniques and products and the territoriality of techniques and local players. It also highlights the processes of encounter, reciprocal influences, power, resistance and negotiation at work, leading to back-and-forth trips and reconfigurations. Particular attention is also given to cultural interaction and the cultural roots of techniques. The analysis of objects or techniques in a given place is followed by their global circulation to understand the influences and evolutions of techniques and products on different continents.

Here, historians of science and technology in China initially helped reinforce this model and then gradually called it into question. Francesca Bray, an anthropologist and historian of agriculture and science and technology in China, followed the path of a number of global products, such as indigo dye, cotton, porcelain and silver metal. This tracking has enabled her to reveal the connections, circulation and transformations of techniques, as well as the cultural interactions that resulted from these circulations. She insists on the need for precise identification of techniques before following their evolution. She shows that, to understand techniques, we need to assess them in their social, cultural, political and economic contexts.

1 This concept of divergence is based on the idea that the economic and commercial trajectories of Europe and China were roughly similar until the mid-18th century, when they began to diverge as the industrial revolution developed.

In an article published in 1998 on domestic techniques, Bray shows, through the analysis of a classic text by a neo-Confucian author, Zhu Xi's *Family Ritual* (1130–1200), that domestic architecture can be considered an important technology in imperial China as industrial design was in the United States in the 19th century (Bray 1998): the ancestral altar, located in the central room of the home, was central to the Confucian sociopolitical order of the time, since Zhu Xi, in his work, had given it a new role not only by imposing ancestral worship on the literate elite and aristocrats but also on all the Chinese and had thus normalized domestic space and standardized social values and practices. Technology is not culture-free, disconnected from culture, but rooted in modes of production that highlight a particular society and its values and largely explain these technologies and their evolution.

This new approach makes it possible to more accurately and subtly trace routes that were hitherto unsuspected, owing to the complexity of the processes at work and the trails that were often difficult to trace: this is the case, for example, with madder dyeing, which originated in India and was imported in several waves between the 16th and 17th centuries, before being learned and then mastered in Europe owing to the help of Armenian dyers who imported and disseminated it there. The proposed approach is based on close observation of the intermediary and the acquisition–hybridization processes centered on techniques.

This model, like the previous one, questions the notion of modernity and the explicit reference this entails to Western history as a frame of reference. This is what Fan Fati, a Taiwanese historian working in the United States, does in his book *British Naturalists in Qing China: Science, Empire and Cultural Encounter*, showing how the British, in their endeavors to discover and categorize China's border regions in the 19th century, imposed their scientific practices (scientific imperialism) on the natives, whom they otherwise depended on for access and residence in these regions (Fan 2004).

This circulation model insists on the need to study techniques in their cultural context, in the society in which they were created and used. This means eliminating concepts and categories imposed on non-Western societies, implying value judgments and implicit hierarchies. These categories appear to be inadequate for assessing these societies and their technologies, as they refer to values that are foreign to these societies, whose objectives differ from those of Western capitalist societies: their vision of man and his place on Earth, their needs and desires, etc. Therefore, it seems necessary to define new concepts and categories that are both neutral (in other words, not implying an explicit or implicit dominant/dominated relationship) and capable of accounting for the desires, values and technologies that these societies have developed.

Among other things, this means rethinking the very idea of technique. In her research, Bray proposed replacing the word *technology* with "material productions and practices that were dominant at the time".

This model was challenged 10 years ago by historians who believed that to understand these circulation phenomena, history should no longer be considered at the national, regional or empire level but rather at the global level.

2.4. Global history model

This model proposes the construction of a history of techniques on a world scale. It goes beyond circulation and aims to develop a historical discourse on techniques on a global scale; to this end, it broadens the scope of the study of circulation.

The analysis is structured on two levels, with intracultural studies at the local level of techniques in context, on the one hand, and an analysis of global connections and their variations, on the other. This new approach thus involves a back-and-forth between the macro and micro levels (Wong 1997): microhistory (through case studies) is structured with a general conceptual framework to enable global connections. Techniques are considered here not only from the point of view of innovation analysis but also, innovatively, from that of knowledge transmission.

By analyzing the role of circulation, mediation and local adaptation (to a local culture and society), this model seeks to highlight differentiation processes and reveal new modes of interaction between people, techniques and cultures: encounters, relationships of domination and resistance, negotiation and reconfiguration in different cultures, or transformations of expertise and objects in new cultures.

This model, following the previous model (see section 2.3), focuses on the epistemological aspect and proposes the pursuit of new methodologies and concepts that are not politically and ideologically marked, refusing to start from traditional Western categories and instead defining new ones that can be applied to all regions of the world in equal terms.

In particular, this model revisits the history of the European Industrial Revolution by analyzing the role played by Asia in the modernization of the production process (Riello 2009; Berg 2015): part of the evolution of production processes that led to the Industrial Revolution can be explained by processes of appropriation and coproduction in England and Northern Europe, which are based on the search for substitutes for imported Asian products (*patkong*, porcelain, tea, printed calico, etc.). This elaboration of a global history of techniques through the study of objects (artifacts) shows that European industrial production owes much to Asian cultures and techniques.

Here again, several historians of Chinese science and technology have helped to reinforce this model. Wong stressed the need to abstract recurrences from case studies (Wong 1997; Berg 2013). In his view, the historian must maintain his research at the local level to master the sources at this level of analysis, but he must then broaden his field of vision beyond national or imperial borders and move up in generality. The aim is, at the global level, to focus on what is common, and at the infra-local level, to distinguish what differentiates these studies from one another. The major difficulty for him is to structure these two levels. To do this, he recommends defining an intermediate level, which presents real methodological difficulty. He suggests using the term "region", even if it remains vague and ill-defined (he gives Fernand Braudel's Mediterranean or Denys Lombard's Javanese space – the South China Sea and South-East Asia – as examples) and refers to entities that are often very diverse in size and therefore difficult to compare. Wong suggested that the region remains a more relevant category than national or imperial spaces but should be used with caution. The latter remains a useful construct for making comparisons and enabling the search for similarities and connections between entities in world history to define concepts and processes that make it possible to explain technical trajectories in different regions of the world, other than through economic development and the technical innovation that underpins it.

Karel Davids, an economic and social historian, has also helped reinforce the global history model. He seeks to explain the divergence in technology between Europe and China from 1500 onward by means other than the price of fuel or labor, and insists on the role of religious institutions and practices in technological development, more specifically in the training, communication and circulation of people and ideas. In his view, the reasons for the industrial take-off of England, and later of Europe, should be sought elsewhere than economic factors. We need to consider variations in religious contexts, in other words, not values and beliefs but rather the impact of religious institutions on society and the role of communications and movements linked to religious traditions, which influence the training and education of populations, the circulation of knowledge and, therefore, technical innovation. In France, for example, he cites the role of religious institutions such as *Les Frères des écoles chrétiennes* (Brothers of the Christian Schools, who provide free basic education), which were heavily involved in the teaching and dissemination of technical knowledge, establishing navigation schools in Vannes and Nantes, a business school in Boulogne and technical schools in other French towns. He also mentions not only the diffusion of technical knowledge that took place within religious orders between their monasteries (Cluniac, Cistercian, etc.) and when new monasteries were created but also the role of pilgrimages in this diffusion and, last, that of missionaries in the circulation of technical knowledge between Europe, the Americas and Asia.

With respect to China, Davids recalls the importance of Buddhism in the creation and development of woodblock printing and the role of Buddhism and Taoism in the development of architecture. In China, as in Europe, the construction sites for religious buildings (temples or cathedrals) were laboratories for innovations that were then applied to secular buildings of all kinds. Similarly, temples and monasteries contributed to the development of new hydraulic techniques, which led to the improvement of certain crops, the development of vineyards and so on. This period of intense construction activity lasted longer in Europe than in China: the relationship between innovation and religion differed between China and Europe from the 14th century onward, when technological development in China took place outside religious institutions, under the aegis of the central state, and individual motivation for innovation therefore disappeared. In Europe, governments, the market and non-religious institutions took precedence over monasteries, but other religious institutions, such as the papacy, took over in the development of innovation, and as the development of patents proves, individual motivation persisted.

Marcus Popplow, a specialist in the history of techniques, also subscribes to the global history model and insists that technology is one of the main factors put forward to explain the great divergence. Currently, we no longer compare scientific knowledge between different countries but look for what constitutes formalized knowledge in different parts of the world (and not just the natural sciences, defined according to the European model).

Popplow asserts that technologies emerged as recognized phenomena and expressed themselves in a highly heterogeneous way over time and space. Innovation changed constantly, as it was defined and delineated by needs and desires that themselves varied in time and space. For him, the consideration given by each era to innovation and technology is the result of complex social and cultural processes. Technology is, above all, a social construct. Research into the history of technical knowledge must be able to integrate the variety of cultural discourses on the different forms of representation of knowledge and ideas, innovative forces and historical choices of technology. These discourses can be identified, in particular, through the study of literary discourse. Specifically, Popplow insists that the relationship between innovations and the written documentation related to them is complex, sometimes even ambiguous, given that no one uses terms such as *innovation* or *technology* in the sense that we understand them today. The interests and activities of governments and elites are often the most visible historical aspect of the spread of technical knowledge. Political leaders and elites have shaped technological developments in lasting and subtle ways through financial and social support rather than through the direct exercise of power.

Finally, Dagmar Schaefer, a historian of science and technology from China, has made a major contribution to developing a new methodology for the global history of techniques. Her starting point is that technology is a global phenomenon since activities such as metallurgy, textile production, agricultural work, water conservation, ceramics production and architecture can be found in numerous cultural spheres and countries. Furthermore, Schaefer noted that technology has developed owing to globalization and seeks a way to approach the history of techniques in the world and traditional Chinese culture while avoiding the pitfalls of Western conceptual categories.

Focusing on Chinese history, she examines how Chinese civilization dealt with technology, how it was conceptualized in writing, and the practices and ideas it helped to develop. This leads us to face different cultures of knowledge. More specifically, she proposes developing an intracultural perspective that enables us to identify the emergence of original concepts related to technology and innovation in the history of this country.

To this end, Schaefer focuses her study on several types of sources. First, she looks at knowledge about objects and their production and how this knowledge is discussed, processed and appropriated. She argues that China developed a specific set of tools for discussing expertise and strategies for recording it, which were found in both material and written culture. She first studies the objects and their productions linked to the central state and the literate elites, as the most visible historical traces of technological development in China are the activities and attitudes of the state and its representatives (a point already mentioned by Popplow).

By analyzing handicraft production, she shows that between the 10th and 19th centuries, Chinese emperors used certain products in their system of governance, treating them as political tools and means of power.

Thus, by appropriating exclusive rights over the production and use of silk or porcelain, monarchs transformed them into tools to maintain or increase their power.

When they were unable to control local production, they imposed rituals to regain a degree of control (particularly over silk and salt).

Schaefer proposed the writings of literati, who were interested in technology, as another source of investigation. Their analysis provides insight into the technical culture that surrounded these scholars and prompted them to write on this subject, traditionally not very popular among their peers.

Schaefer also proposed analyzing the archives of the administrations responsible for the production of technical objects. In particular, she has begun systematic research into the seals that were attached to these objects. The study of these seals enables us to understand the changes in ideological support that emperors gave to certain technologies and then others. In particular, the study of handcrafted seals allows us to make a link between objects, their producers and their owners. The study of porcelain and lacquer seals provides a glimpse into the variety of emperors' attitudes in their appropriation of practical and technical knowledge and sheds further light on evolutions in political control and relations between local and central levels, which written sources do not reveal.

Finally, Schaefer is interested in steles erected by officials, individuals or guilds and placed at the entrances to factories, workshops, temples, bridges or other places linked to an activity involving techniques. These steles provide invaluable information on the importance given to a particular production and function as information panels, not only making it possible to follow the spatial evolution of technical developments but also, in the case of steles representing associations of craftsmen, such as paper or lacquer manufacturers and blacksmiths, to learn about their modes of organization, their rules, production processes and so on. They reveal the different ways in which technologies and their innovations are appropriated at the local or individual level and show that, in an empire under centralized power, different cultures of knowledge coexist, particularly in the technological field, serving several functions at once.

2.5. Conclusion

The revolution that the history of techniques has undergone in the space of several decades has been remarkable: tools, methods, objects and questions have been transformed, enabling us to move from the study of major inventions and their inventors, centered on Western Europe, to research on technology in the broadest sense, the causes and consequences of its appearance, its evolution and its environment, and the political, but above all, cultural and societal structures that underpin it.

Nowadays, there is an emphasis on a global vision, which is based on local studies that account for the circulation of objects, people and ideas, shedding as much light as possible on the processes of transformation and reconfiguration that this entails at a global level. To achieve this, new questions, concepts and sources are used or called upon, allowing for greater reflexivity and backtracking to describe and understand the complexity at work in a global world. Studies on the history of techniques in China, which are increasingly numerous and varied, play an active part in these developments.

2.6. References

Berg, M. (ed.) (2013). *Writing the History of the Global: Challenges for the 21st Century*. Oxford University Press, Oxford.

Berg, M. (2015). The merest shadows of a commodity: Indian muslins for European markets 1750-1800. In *Good from the East, 1600-1800 Trading Eurasia*, Berg, M. (ed.). Palgrave MacMillan, Basingstoke.

Bray, F. (1998). Technics and civilization in late imperial China: An essay in the cultural history of technology. *Osiris*, 13, 11–33.

Davids, K. (2013). *Religion, Technology and the Great and Little Divergence. China and Europe Compared.* Brill, Leiden.

Davids, K. (2016). Religion and technological development in China and Europe between about 700 and 1800. *Artefact. Techniques, histoire et sciences humaines*, 4, 287–306.

Fan, F. (2004). *British Naturalists in Qing China: Science, Empire and Cultural Encounter*. Harvard University Press, Cambridge.

Pomeranz, K. (2000). *The Great Divergence: China, Europe, and the Making of the Modern World Economy*. Princeton University Press, Princeton.

Popplow, M. (2016). Technology and technical knowledge in the debate on the 'great divergence'. *Artefact. Techniques, histoire et sciences humaines*, 4, 275–286.

Riello, G. and Tirthankar, R. (eds) (2009). *How India Clothed the World: The World of South Asian Textiles, 1500-1850*. Oxford University Press, New York.

Rosenberg, N. (1970). Economic development and the transfer of technology: Some historical perspectives. *Technology and Culture*, 11, 550–575.

Rosenthal, J.-L. and Wong, R.B. (2011). *Before and Beyond Divergence: The Politics of Economic Change in China and Europe*. Harvard University Press, Cambridge.

Schäfer, D. (2013). Technology and innovation in global history and in the history of the global. In *Writing the History of the Global: Challenges for the 21st Century*, Berg, M. (ed.). Oxford University Press, Oxford.

Sivin, N. (1982). Why the scientific revolution did not take place in China – Or didn't it? *Chinese Science*, 5, 45–66.

Wong, R.B. (1997). *China Transformed: Historical Change and the Limits of European Experience*. Cornell University Press, Ithaca/London.

Wong, R.B. (2013). Regions and global history. In *Writing the History of the Global: Challenges for the 21st Century*, Berg, M. (ed.). Oxford University Press, Oxford.

3

Invention: Rehabilitating a Concept

Liliane HILAIRE-PÉREZ
EHESS, Université Paris Cité, Paris, France

3.1. Introduction

In the 1990s, as the history of techniques began to re-enter historical research, David Edgerton challenged the importance of innovation in this discipline. The sociologist believed that:

> [...] most writings on the history of techniques and the relationship between techniques and society deal with innovation and the emergence of new techniques. They do not distinguish these topics from the study of techniques that are necessarily older and have become widely integrated into society, and which are often considered by historians of techniques to be outdated, obsolete, or residual, and therefore of little interest. (Edgerton 1998, p. 816)

This critical stance on the role of invention and innovation in the history of techniques was shared by many historians in France. Denis Woronoff encouraged us to view the history of techniques as something far removed from success stories and as a history marked by slow change, by incompleteness, by the coexistence rather than succession of technical generations in a world of work largely dominated by repetition, and by the mastery of skills acquired over time in a world where innovation did not always offer comparative advantages (Woronoff 1998). The development of techniques in society and their validation by human communities

History of Techniques,
coordinated by Guillaume CARNINO and Xavier GUCHET.

became essential parameters for historians, as Catherine Verna explained (Verna 2002).

At the same time, Serge Benoit, whose work has just been compiled (Benoit 2020), showed that industrialization in France in the 19th century had undergone several phases and that the long term was also a scale to be considered in innovation processes, which were too often perceived as sudden accelerations and radical transformations. Technical invention covered a variety of phenomena that had been masked by the dominance of the notion of the industrial revolution, as well as by the sacralization of progress and mechanization since the 19th century and then by the cult of innovation, growth, development and the prowess of "technology" after World War II, which the history of techniques has now begun to analyze (Leenhardt 2021; Neumann 2022).

The scope of work on invention and innovation over the past generation is a legacy of these reinterpretations. However, unlike Edgerton's critique, which tended to replace the history of uses for that of innovation, the vast body of research that developed at that time made it possible to reopen the question of invention in light of continuist approaches focusing on the durability of techniques and the reuse of objects and processes that go hand in hand with skills of adjustment, substitution and transposition. It has now become clear to historians that the opposition between invention and use is inconsistent, as studies have shown not only the interference between technical generations and the coexistence of multiple temporalities in all periods but also the incremental nature of invention, which lies at the very heart of the most mundane and seemingly "static" activities, such as maintenance and repair (Bernasconi et al. 2022). These overlapping and extended periods of invention, which are based on capacities for adaptation and improvement rooted in an analogical and synthetic mode of thinking (Hilaire-Pérez 2017), also depend on the role of omissions, resurgence and latent circulation, an essential aspect of inventive processes, which have been seen for centuries as an art of memory (Vérin 1993).

It is therefore a question of understanding the concept of invention outside the teleologies of progress, in light of questions about the meanings given to invention by actors, about "regimes of historicity" and the experience of time in the past, about intellectual processes and the place of invention in working communities, in line with current practices, and not myths about growth. This does not mean that representations of invention are excluded from the field of research. The study of myths of progress and utopias is central to understanding how society has invested in invention, as well as the political and social issues that have shaped it. This was the focus of Christine MacLeod's seminal study on the heroization of the inventor in

Victorian England before scientists and industrialists took over as heralds of modernity (MacLeod 2007).

Owing to this trend, which has made it possible to historicize representations of invention, an in-depth analysis of inventive practices has been developed. We present these new areas of research. We will focus on the modern period, which generated many inventions in fields that were by no means limited to industrialization, but we also delve into the contemporary period, which has seen inventive activity transformed by the rise of internalized research.

3.2. Technical intelligence in question

The intellectual processes inherent in all techniques are at the heart of this renewal of studies on technical invention, whose history has long been reduced to causes endogenous to production sectors and a linear progression of human progress in a Europe-centric view. Hélène Vérin's book on engineers in modern Europe was a crucial step forward: it dissected *ingenium*, genius and the ability to resolve "universal contradiction" (Vérin 1993). Both a philosopher and historian, the author highlighted the synthetic, analogical, substitutive and combinatorial thinking inherent in invention, conceived as the search for "middle ground". Pascal Brioist's analysis of Leonardo da Vinci's inventing echoes this understanding of genius as comparative and synthetic thinking, born of an observation of reality and workshop work (Brioist 2019). It is from experimentation that da Vinci abstracted theorization and thought of the world as a system.

Pascal Brioist also joins Hélène Vérin's fundamental study on reductions in art, a literary genre that aims to theorize invention by formalizing the principles of action that must be arranged, combined and adapted to concrete situations (Dubourg-Glatigny and Vérin 2008). This "mechanization" of invention, designed to facilitate action "in the miscellaneous", involves a profusion of works that highlight the fruitfulness of the mind and its ability to generate solutions, such as machine theaters (Dolza and Vérin 2004). In *Le Diverses et artificiose machine* (1570), engineer Agostino Ramelli demonstrated the human mind's ability to produce multiple effects through the proliferation of inventive tricks on a hundred pumps, expressing the power of the mind to constrain nature (Vérin 2002). While their authors did not rule out the possibility of practical applications for these inventions, the multiple ways of constructing them and the descriptions of the effects produced are valuable expressions of technical ingenuity.

These richly illustrated collections fuel imagination with possibilities. They also highlight the intellectual and liberal nature of invention, which cannot be assimilated into the mechanical arts. In these treatises, the capacity for invention suggests a breakdown of barriers between different trades. The synthetic thinking that underlies invention, as well as the analogies between different types of manufacturing, presupposes the establishment of a "technical space", as it is called by Hélène Vérin, which cannot be reduced to categories of trades but also to any principle of applying science to practice. This was also the creed of encyclopedists, notably Denis Diderot, who called on artisans to collaborate and exchange ideas through workshop visits and a Baconian reform of the arts and crafts. For Diderot, invention is the result of transpositions and adaptations (Hilaire-Pérez 2002). To invent is to discover "useful and unknown relationships" between the arts and between the arts and the natural world – "connections" that humans, too distant from one another, are unable to uncover. Diderot expressed the wish that artisans compare their processes and lend each other "mutual aid" under the aegis of an "artisan's monarch".

The motif of the artisan, who is embedded in specific trades, lacking knowledge of his art beyond his own practice, resistant to innovation and to any opening up of knowledge, is short-sighted in the long term. The socioeconomic context is conducive to the development of a scholarly formalization of craft activities, which is essential for the rise of the nation-state and entrepreneurship. In this economic and theoretical space, an encyclopedic knowledge of arts and crafts developed, postulating the need for artisans to be enlightened by scholarly knowledge, which alone would ensure the circulation of knowledge and invention by encouraging manufacturers to borrow and transpose solutions from one manufacturing process to another. This was one of the challenges of the science of arts or technology, as a branch of cameralist knowledge developed in the German world. One of its experts, Johann Beckmann, published *Entwurf der allgemeinen Technologie* (Carnino et al. 2017) in 1806. He explains that "many trades, however different their materials and products, have to perform various operations with the same aim in mind, or know how to achieve the same aim in very different ways". This was the outline of a general theory of operations that was applicable across trades. The aim was to guide artisans toward innovative transpositions: "This would help artists and artisans to acquire general and solid knowledge of the objects they work on and the usual processes to be followed and to guarantee them at least a direction that could guide inventive minds toward new and useful improvements".

The technological model of invention underpins an elitist representation. Although invention owes nothing to process science, which is still so inconclusive

(Belhoste 2011), it is not considered to be the product of artisanal practices. It is supposed to come from experts and artisans enlightened by science, who have acquired the codes of the elite and seek to distinguish themselves with the title of "artists" (Bertucci 2017; Nègre 2017). The breaking down of barriers between the arts and crafts was promoted by the learned elites, who sought to describe and codify them as part of a general movement to standardize the discourse and representations of invention, as in patents (Baudry 2018; Guffroy 2023).

Is this not a form of appropriation, of scholarly and entrepreneurial control of practical knowledge? This is what several studies suggest. For Daniel Fischer, who studied the metallurgist Frédéric de Dietrich in the service of the royal administration and its innovation policy, craft techniques are "cross-disciplinary knowledge" whose mastery underpins scholarly legitimacy (Fischer 2022). The quasi-anthropological description of actions is a major resource for men of science who, until the end of the 19th century, remained dependent on knowledge of craftsmanship, both in the arts and in industry (Barron 2022). In metallurgy, for scientists and engineers, invention has long been a matter of observing – and capturing – techniques in situ (Belhoste 2004; Cantelaube 2009). The codification of the arts appeared to be the most secure way to improve processes. This is also suggested by a recent study of English agronomic literature, whose authors appropriated vernacular techniques and turned them into a tool for the innovative management of large estates (Fisher 2022).

However, Hélène Vérin emphasized the existence of a fundamental tension between the advantages of reductions in art and knowledge gained from experience, noting that novelty is based on chance, on the singularity of action in experience, on defining rationality as a sensitive commitment and the implementation of "innate intellectual faculties" to "deal with things", in short, on the "engine", an empiricism "guiding design in diversity" (Vérin 1998). Is the act of knowledge itself not compromised by the enthusiasm for treatises that accompanied the rise of publishing in Europe? Hélène Vérin noted that, faced with the "mechanization" of invention, Francis Bacon was wary of such an approach, which "prematurely reduces knowledge to arts and methodical presentations" and "puts an end to questions" ("the overearly and peremptory reduction of knowledge into arts and methods") (Vérin 2008). The denunciation of the spirit of the system has continued to spread, pointing to the growing antinomy between "codification and invention" and "determination and fertility". It is toward this empiricism, down to the most mundane, but never trivial, workshop practices, that research is currently focused.

3.3. Banality of invention

In the 1980s, philosopher Judith Schlanger proposed reflecting on intellectual invention as inseparable from the notion of understanding, defining the act of thinking by its relation to "the mass, which grounds it as a meaningful gap". She suggested the importance of "the ordinary experience of knowing thought, [...] that which is mundane and leaves no trace" and yet "can be described in the same terms as the experience of extraordinary thought", since "ordinary reflection and the treasure of thought have in common" a quality, that of "establishing a communicable difference" (Schlanger 1983, pp. 246–247). Far from confusing any hierarchy between works of the mind, or even between all human productions (intellectual or material), her point of view encouraged investigation into the banality at the heart of invention, repetition, common and shared thought, and the processes of reproduction, transmission and learning. More than the intellectual gesture, it was the quality of the act of thinking that Judith Schlanger questioned, the nature of displacement, whether it be common or creative, seen as a "related disengagement". The philosopher invited us to consider that even the most mundane reflection, embedded in the heart of the "understanding of what has already been said" and the "repetition of almost the same thing" ("the two sides of the inventive minimum"), was worthy of interest (Schlanger 1983, p. 243).

This reflection resonated with the history of material culture and cultural practices, as seen in the work of Daniel Roche, a reader and promoter of the philosopher's work, who transposed these concepts from the history of the intelligible to that of the sensible. This approach to banality was in line with (and supported) the study of civilizations through "ordinary practices and the place of objects in everyday life", inseparable from the choices and aptitudes for change that emerged at the very heart of repetition, as shown in *Le peuple de Paris*, then in *L'histoire des choses banales* (Roche 1993, 1997). The theme was essential to the history of consumption and to that of techniques, which Fernand Braudel had long defined as "structures of everyday life" (Braudel 1979). The mundane, the commonplace, present to all and yet absent from history, was in a way the hallmark of techniques in history, as the founders of the *Annales* school reminded us (Bloch 1935). The history of invention was no exception to the long duration and structures of everyday life.

In the proceedings of the symposium on *Les archives de l'invention* ("The archives of invention"), we asked: "How can we avoid a managerial and one-dimensional approach to invention that erases the plurality of actors, the collectives of invention, uncertainty, failures, and trial and error in research? How can we measure exchanges, collaborations, the role of networks within companies, and the contribution of workshop invention?" (Hilaire-Pérez 2007). As we have said, few writings are capable of capturing tacit knowledge, even in the contemporary era of R&D. Invention archives fall under the evidential paradigm, consisting of tenuous traces of practices that are complex to identify and that are often erased owing to the very banality of inventive activity, whereas the preservation of laboratory writings has been a fundamental part of the research process. With regard to the chemist Pierre Potier, Muriel Le Roux explained that "the numerous notes he took during his observations, which fed his intuitions and reflections, were archived" (Le Roux 2007). The eagerness of historians of invention to multiply their efforts to preserve, protect and publish traces of inventive activity is understandable, given that invention is in itself evanescent, doomed to absence and oblivion (Ratcliff 2022). This is also why studies focus on "the epistemological role of narrative in the innovation process" as a means of rationalizing and legitimizing an exploratory, uncertain and indecisive activity (Chouteau et al. 2022).

This approach encouraged us to question the meaning of the word *invention* and to revise its nature of absolute novelty. Such distancing is essential, just as in past societies, where the term did not have the same meaning as it does today and was therefore associated with imitation, repetition and the ability to connect facts and construct relationships between distant objects, and between nature and living beings (Vérin 1993, 2008; Hilaire-Pérez 2002, 2013; Lanoë 2008, 2021; Brioist 2019; Thébaud-Sorger 2020; Baudry et al. 2022). In *L'Histoire et secret de la peinture en cire* (1755), Denis Diderot argues that invention is the art of discovering "unknown relationships" and calls on artisans to lend each other "mutual aid" (Hilaire-Pérez 2002). The place of synthetic and analogical thinking, as identified by Hélène Vérin, has also mobilized historians of the early periods, for whom identifying the inventive process is a real challenge (Verna 2000). However, it is in the English case that the subject was most strongly revisited. While England represented the textbook case for the notion of industrial revolution and radical inventions, it was the site of a number of reinterpretations in favor of a broader definition of the concept of invention, including product innovation (models, materials, dyes, sparkles), improvements in tools and marketing techniques. In these reinterpretations, invention was also seen as an exercise in taste, the pleasures of the imagination and the importance of imitation as an aesthetic principle, as theorized by Adam Smith in 1777 (*On the Nature of that Imitation which takes Place in what are called the Imitative Arts*) (Berg 2002; Hilaire-Pérez 2013; Guffroy 2023).

Invention emerged as a variation of use, as an adaptation to specific needs, as confirmed by current research on repairs (Fennetaux 2022; Pautet 2022). In addition to opposing innovation and usage, technical historians have revisited the role of users, including them in the cycle of invention through their ability to appropriate processes and demonstrate practical intelligence (Hilaire-Pérez and Thébaud-Sorger 2006). This is why the history of technology as a public culture now intersects with that of invention. The Enlightenment era is particularly well studied. It is characterized by the solicitation of the public's ability to debate the advantages of techniques and propose inventions, owing to the increasing availability of useful knowledge. Newspapers, particularly the advertising press, gave rise to genuine communities of text, networks of correspondents, readers, subscribers, booksellers and advertisers who appropriated the medium, built social networks and turned their journals into a collective project. As Marie Thébaud-Sorger has shown in the case of the aerostation, and more broadly for the many devices for capturing air, heating and ventilation, the press, how-to books, public demonstrations and the market for scale models gave rise to hundreds of proposals and revealed a shared, collective invention made up of borrowings, circulations and transpositions. Among scholars, artisans and amateurs working with gases and fluids, a cross-disciplinary and practical understanding of appropriate construction processes has developed, leading to a profusion of inventions to ensure the tightness, resistance, capture and circulation of these substances (Thébaud-Sorger 2018, 2020).

It is in light of these cross-disciplinary connections that we are gaining a better understanding of the inventive process. The matrix is the workshop, a genuine crossroads of techniques and social interactions (Lanoë 2024), whose transductive function can be observed over the long term, as in the laboratory that emerged from it and has become the preferred space for research (Bret and Lanoë 2006). The banality of invention, the link with imitation and the social dynamics of collaboration characterize design practices from the world of the artisan workshop to that of the R&D laboratory. Considerable research has highlighted the inventive work carried out by a wide variety of actors, in both craftsmanship and industry, at the very heart of productive structures.

3.4. Inventive labor

Invention is part of everyday work. In terms of the workshop, it is based on close, intimate knowledge of materials, which allows for endless combinations, adaptations and transpositions.

In 18th-century London, the daily grind of repair work in the coachbuilding trade was productive ground for the inventive culture of mechanics. The Joseph Jacob father and son family is a case in point, owing to archives spanning two generations (Hilaire-Pérez 2013). Joseph Jacob Jr. was a prominent coachbuilder, inventor, patent holder and correspondent for the Society of Arts, for which he wrote a treatise on traction. He started working with his father, whose accounts show repeated repairs in the 1720s, which were carried out in a series of production steps. For example, to attach a new slat to the roof of a sedan chair, his father glued the canvas, sewed the lining and nailed the leather. Although we cannot know whether he ever substituted one action for another within a work sequence, such as sewing instead of gluing, we can see that he noted substitutions of materials: on two occasions, a thin yellow canvas was used "in place" of a window pane, or a door window was recut for a front panel. In Joseph Jacob Sr.'s business, the economics of repair gave rise to synthesis practices, which were based on combining similar actions and substituting materials. This analog and synthetic technical culture nurtured inventiveness and transposition to the next generation. In Joseph Jacob Jr.'s work, practical intelligence became more complex. The son excelled in mechanical devices and their multiple applications. In 1774, he took out a patent for removable carriage ornaments that could be adapted to buildings, furniture and musical instruments. Another patent taken out by Joseph Jacob in 1803 involved axle boxes that could be adapted to various machines. Carriage building thus created a "technical space" (Vérin 1993). Even before the rise of mechanization, and as the intermediate sector of repair techniques developed, emerging cross-cutting techniques were shared, structuring trades around operational sequences and underpinning inventiveness in this sector.

These synthesis operations, which are key to inventiveness, are all the more developed when artisans participate in working communities, networks and mobility that favor intellectual processes of composition, transposition, borrowing and analogy (Lanoë 2008; Hilaire-Pérez 2013). The relational nature of invention and its links with collaborative skills, exchange and movement (in all senses of the term) are one of the contributions of recent studies that highlight the decisive role of invention collectives and open technique (Foray and Hilaire-Pérez 2005). Pamela H. Smith has shown that since the Renaissance, goldsmiths and painters have combined knowledge and multiple activities and composed their products with the help of multiple contributors, both in the workshop and networks, to meet growing demand (Smith 2004).

Working communities constantly borrow techniques from other workshops and trades. Lyon, the most inventive city in France under the Old Regime, is a case in point. It was a genuine hotbed of collective invention, supported by the municipality and the Grande Fabrique (the guild of gold and silver silk weavers), which encouraged inventiveness through widely distributed bonuses, promoting multiple improvements in trades using programmed pull-string machines with rollers or perforated cards – a system that benefited Jacques Vaucanson (1709–1782) and Joseph Marie Jacquard (1752–1834) (Hilaire-Pérez 2008). At all levels, the system encouraged networking between manufacturers. Local exchanges were easily identifiable owing to distribution bonuses financed by the municipality. In addition to encouraging the dissemination of inventions, these relationships were also a source of emulation and competition. Other links emerged from the files of inventors in the silk industry, with carpenters, locksmiths and foundry workers; inventive mobilization arose from the growth of a mechanical sector. While it is difficult to assess the economic effectiveness of this system, it can be assumed that it promoted "collective invention", which was reinforced by clauses requiring models to be deposited at the guild's premises and by obligations to teach the use of new techniques.

Other transfers, outside any institutional framework, mark the everyday life of the workshop, as can be seen in the archives of the practice. Recent work shows that craft techniques are governed by two principles: contiguity (trades linked by proximity and complementarity) and analogy (trades linked by common operations). Technologists identify and formalize new operational concepts (Mertens 2002) as common practices that are perfectly mastered by artisans (Hilaire-Pérez and Lanoë 2011). This structure underlies Catherine Lanoë's analysis of artisanal inventiveness. Studying glove makers and perfumers, for example, she shows "that the experience and empirical knowledge of the characteristics of the raw material acquired in the manufacture of one product enables the glove maker and perfumer to think extensively about its various properties in the composition of another product" (Lanoë 2021).

On the one hand, contiguity relationships exist, as she explained for the creation of new dyes in glove-making in the 18th century based on borrowings from textile dyeing (Lanoë 2021); thus, "the 1699 edition of Simon Barbe's manual includes a new section entitled 'How to dye goat skins in several colors'. This new terminology is accompanied by new operations, such as the preliminary annealing of the hides and the preparation of various color baths [...] transposed from the art of textile dyeing. Perhaps drawing on his travels in France and Italy, and a collection of processes gathered from various glove makers, [...] Simon Barbe was familiar with and encouraged these transpositions". For Catherine Lanoë, "well before the years 1730–1750, marked at the Academy of Sciences by a great effervescence around the art of dyeing, [...] glove makers and perfumers applied themselves to innovation, transposing and adapting the knowledge built up in this other sector of craftsmanship to their own ends".

On the other hand, this type of borrowing from related professions is coupled with analogous cross-fertilization with distant professions, whose processes are transposed with the aim of inventing new devices (Lanoë 2008). The author again draws on practical examples:

> The probate inventories of Parisian glove makers and perfumers reveal that, in order to increase powder production, these artisans mechanized the sifting process by transferring it. From the 1740s onwards, sieving machines (or the French *bluteau*) were copied from millers and bakers, whom they frequented for supplies. Glove makers and perfumers identified a similarity and adapted the technique. Thus, the operational abstraction behind the invention belongs entirely to the practitioners.

Recent research has therefore provided a better understanding of this model and its two organizing principles – cross-functionality and contiguity – which had unparalleled structural force in the 18th century in connection with the rise of subcontracting. The lengthening of production chains – a widespread practice in all trades – reinforced the importance of assembly, substitution and composition techniques, which formed the bulk of artisanal work.

Here too, the differences between the worlds of craftsmanship and emerging industry are becoming increasingly pronounced. Their common ground lies in collective intelligence, the driving force behind any type of production. The very concept of collective invention was developed by Robert C. Allen based on his study of information sharing among English steel companies in the 19th century (Allen 1983). There is no doubt that the rise of large companies and economic concentration, as well as applied science and the internalization of research, have diminished the role of knowledge collectives.

However, much work on the relationship between science and industry in the 20th century has shown that internalized research is also based on "a plurality of modes of production and acquisition of knowledge" (Galvez-Behar 2008, p. 145, 2020). While the laboratory eventually became the norm, interprofessional collaborations developed, as identified by Peter Galison, who saw microelectronics at MIT's Rad Lab as the result of local coordination between subcultures ("experimentation, instrument-making and theory") (Galison 1997, p. 46). Since then, several studies have confirmed the complexity of internalized research. Gabriel Galvez-Behar highlighted the importance of shop-floor invention and its regulation through patents, Hélène Mialet described the "distributed" identity of the inventor within the company and Muriel Le Roux highlighted the collectives of engineers in the aluminum industry, as well as those of researchers, chemists, academics and industrialists in the pharmaceutical industry (Mialet 2008; Le Roux 2013). Muriel Le Roux said the following about Pierre Potier's archives: "The higher up he climbs in the hierarchy, the more you have to read his notebooks alongside those of his colleagues who carried out the work for him. For a complete study, you would need to analyze his colleagues' notebooks at the same time as his own" (Le Roux 2007). Sylvie Delvenne and Didier Devriese also noted that R&D is inextricably linked with "tricks" and "verbal knowledge", which are often forgotten, as is the case with "all innovations that have been industrialized": "the laboratory notebook only gives an imperfect account of the interactions taking place between the various scientists (nor does it give a better account of the interactions between scientists and technicians, or [...] of the interactions between technicians themselves)" (Delvenne and Devriese 2007). It is clear that R&D relies on cooperation between experts within a company, integrating "innovation workers" in various ways (Cloutier 2014; Boisard et al. 2016), while creating links outside the company as scientists become involved in corporate research (Mercelis et al. 2017; Galvez-Behar 2020).

This is why, today, historians are taking a critical look at the standardization of interprofessional relations through contract-based project research, which weakens the potential for hybridization of knowledge (Lanciano-Morandat 2019). A comment by Muriel Le Roux gives food for thought: "Even if we can say that scientific research has replaced the multiple sources of invention, many companies find themselves at an impasse because they have discarded the artisanal – even informal – approach to innovation, giving little space to local knowledge, craftsmanship, and learning through trial and error" (Le Roux 2018). In short, the collective dynamics of invention constitute a genuine model, despite being repeatedly called into question.

We can see how the history of invention has been at the heart of reinterpretations and renewals of the history of techniques for a generation. Not only has it been at the forefront of debunking progress, but it has also helped to develop an intellectual history of techniques and to restore the mental and cognitive processes that underpin the search for solutions, particularly in the mundane, but never trivial, everyday exercise of work.

One might be tempted to say that the historicization of invention, far from the grand narratives of the industrial revolution, the sacralization of engineers, growth and innovation, gives new substance to the history of technical culture. As sociologists have noted, "the culture of innovation [has] obscured technical culture" (Chouteau et al. 2017). In this respect, the 18th century occupied a significant place as an observatory of this historiographical revival. Contrary to a vertical conception of invention as the result of interactions between scholars and practitioners, as conveyed by the concepts of the Industrial Enlightenment and the Artisanal Enlightenment (Mokyr 2009; Bertucci 2017), it is clear that for contemporaries, inventive resources were found, on the one hand, in workshops, in a heterogeneous, transductive technical culture that cannot be reduced to any codification, and, on the other hand, in the past and its resurgences, in the long history of techniques – the common heritage of humanity discovered by 18th-century archaeologists (Montègre 2024). Although the word "technique" as a noun was not used at that time (Camolezi 2022), this is indeed what it refers to. Neither art nor science invention is the manifestation of the power of the mind.

3.5. References

Allen, R.C. (1983). Collective invention. *Journal of Economic Behaviour and Organization*, 4, 1–24.

Barron, G. (2022). La technologie en France au XIXe siècle : de la pratique au concept au travers du parcours d'un acteur. *Artefact*, 15(1), 153–174.

Baudry, J. (2018). Écrire et dessiner l'invention : les brevets et la technologie en France et aux États-Unis. In *La technologie générale: Johann Beckmann Entwurf der algemeinen Technologie/Projet de technologie générale (1806)*, Carnino, G., Hilaire-Pérez, L., Hoock, J. (eds). PUR, Rennes.

Baudry, J., Blanc, J., Hilaire-Pérez, L., Ratcliff, M., Wenger, S. (eds) (2022). *Produire du nouveau ? Arts – techniques – sciences en Europe (1400-1900)*. Éditions du CNRS, Paris.

Belhoste, J.-F. (2004). Pourquoi Frédéric Le Play, métallurgie, explorateur de la Russie, s'intéressa à la question sociale. In *Autour de l'industrie : histoire et patrimoine. Mélanges offerts à Denis Woronoff*, Belhoste, J.-F., Benoit, S., Chassagne, S., Mioche, P. (eds). CHEFF, Paris.

Belhoste, J.-F. (2011). La question de l'acier en France au XVIIIe siècle : l'histoire d'un rendez-vous manqué entre science et industrie. In *L'acier en Europe avant Bessemer*, Dillmann, P., Hilaire-Pérez, L., Verna, C. (eds). Presses universitaires du Midi, Toulouse.

Benoit, S. (2020). *D'eau et de feu : forges et énergie hydraulique, XVIIIe-XXe siècle. Une histoire singulière de l'industrialisation française*. PUR, Rennes.

Berg, M. (2002). From imitation to invention: Creating commodities in the eighteenth century. *Economic History Review*, 55(1), 1–30.

Bernasconi, G., Carnino, G., Hilaire-Pérez, L., Raveux, O. (eds) (2022). *Les réparations dans l'Histoire. Cultures techniques et savoir-faire dans la longue durée*. Presses des Mines, Paris.

Bertucci, P. (2017). *Artisanal Enlightenment: Science and the Mechanical Arts in Old Regime France*. Yale University Press, New Haven.

Bloch, M. (1935), Les "inventions" médiévales. *Annales d'histoire économique et sociale*, 7, 634–644.

Boisard, P., Didry, C., Younès, D. (eds) (2016). *Les travailleurs de l'innovation. De l'entrepreneur aux salariés.* PUR, Rennes.

Braudel, F. (1979). *Civilisation matérielle, économie et capitalisme, XVe-XVIIIe siècle 1 : les structures du quotidien.* Armand Colin, Paris.

Bret, P. and Lanoë, C. (2006). La formation d'un espace de travail entre sciences et arts et métiers : le laboratoire du chimiste et le laboratoire du parfumeur au XVIIIe siècle. In *Le travail avant la révolution industrielle.* CTHS, Paris.

Brioist, P. (2019). *Les audaces de Léonard de Vinci.* Stock, Paris.

Camolezi, M. (2022). Technique, Technics, Technik, substantifs : les mots à travers les dictionnaires du XIXe siècle. *Artefact*, 15, 61–106.

Cantelaube, J. (2005). *La forge à la catalane dans les Pyrénées ariégeoises, une industrie à la montagne (xviie-xixe siècle).* CNRS/Université de Toulouse-Le Mirail, Toulouse.

Cantelaube, J. (2009). Les Annales des mines et la forge à la catalane. In *La presse et les périodiques techniques en Europe (1750-1950)*, Bret, P., Chatzis, K., Hilaire-Pérez, L. (eds). L'Harmattan, Paris.

Carnino, G. (2015). *L'invention de la science. La nouvelle religion de l'âge industriel.* Le Seuil, Paris.

Carnino, G., Hilaire-Perez, L. and Hoock, J. (eds) (2017). *La Technologie générale. Johann Beckmann Entwurf der allgemeinen Technologie / Projet de technologie générale (1806).* PUR, Rennes.

Chouteau, M., Forest, J., Nguyen, C. (2017). Quand la culture d'innovation fait écran à la culture technique. Technologie et innovation [Online]. Available at: https://www.openscience.fr/Quand-la-culture-d-innovation-fait-ecran-a-la-culture-technique.

Chouteau, M., Forest, J., Nguyen, C. (eds) (2022). Mises en récit d'innovations. *Artefact. Techniques, histoire et sciences humaines*, 16, 9–129.

Clifford, H. (1999). Concepts of invention, identity and imitation in the London and provincial metal-working trades, 1750-1800. *Journal of Design History*, 12(3), 241–255.

Cloutier, L. (2014). Inventeurs entrepreneurs. In *Dictionnaire sociologique de l'entrepreneuriat*, Chauvin, P.-M., Grossetti, M., Zalio, P.-P. (eds). Presses de Sciences Po, Paris.

Delvenne, S. and Devriese, D. (2007). Archives de l'invention et sources orales. Pour l'emploi de nouvelles méthodologies en histoire des processus de création. In *Les archives de l'invention. Ecrits, objets et images de l'activité inventive des origines à nos jours*, Corcy, M.-S., Demeulenaere-Douyère, C., Hilaire-Pérez, L. (eds). CNRS, Toulouse.

Dolza, L. and Vérin, H. (2004). Figurer la mécanique : l'énigme des théâtres de machines de la Renaissance. *Revue d'histoire moderne et contemporaine*, 51(2), 7–37.

Dubourg-Glatigny, P. and Vérin, H. (2008). *La réduction en art, un phénomène culturel.* In *Réduire en art : la technologie de la Renaissance aux Lumières*, Dubourg-Glatigny, P. and Vérin, H. (eds). Éditions de la Maison des sciences de l'homme, Paris.

Edgerton, D. (1998). De l'innovation aux usages. Dix thèses éclectiques sur l'histoire des techniques. *Annales. Histoire, sciences sociales*, 4(5), 815–837.

Fennetaux, A. (2022). Consommation industrieuse: réparations textiles, raccommodages et faire durer en Grande-Bretagne aux XVIIIe et XIXe siècles. In *Les réparations dans l'histoire. Cultures techniques et savoir-faire dans la longue durée,* Bernasconi, G., Carnino, G., Hilaire-Pérez, L. and Raveux, O. (eds). Presses des Mines, Paris.

Fischer, D. (2022). *Philippe Frédéric de Dietrich. Un entrepreneur des savoirs au XVIIIe siècle*. Presses des Mines, Paris.

Fisher, J.D. (2022). *The Enclosure of Knowledge: Books, Power and Agrarian Capitalism in Britain, 1660-1800.* Cambridge University Press, Cambridge.

Foray, D. and Hilaire-Pérez, L. (2005). The economics of open technology: Collective organization and individual claims in the "fabrique lyonnaise" during the Old Regime. In *Frontiers in the Economics of Innovation and New Technology. Essays in Honor of Paul A. David*, Antonelli, C., Foray, D., Hall, B.H., Steinmueller, E.W. (eds). Edward Elgar Publishing, Cheltenham.

Galison, P. (1997). *Image and Logic: A Material Culture of Microphysics.* University of Chicago Press, Chicago.

Galvez-Behar, G. (2008). *La République des inventeurs. Propriété et organisation de l'innovation en France (1791-1922).* PUR, Rennes.

Galvez-Behar, G. (2020). *Posséder la science. La propriété scientifique au temps du capitalisme industriel.* Édition de l'EHESS, Paris.

Goulet, F. and Vinck, D. (eds) (2022). *Faire sans, faire avec moins. Les nouveaux horizons de l'innovation.* Presses des Mines, Paris.

Guffroy, Y. (2023). Représenter l'invention : étude de l'évolution du dessin d'objet technique en Angleterre (ca. 1750-1850). PhD Thesis, École polytechnique fédérale de Lausanne, Lausanne.

Hilaire-Pérez, L. (2000). *L'invention technique au siècle des Lumières.* Albin Michel, Paris.

Hilaire-Pérez, L. (2002). Diderot's views on artists' and inventors' rights: Invention, imitation and reputation. *British Journal for the History of Science*, 35, 129–150.

Hilaire-Pérez, L. (2007). Technology as a public culture in the XVIIIth century : The artisans' legacy. *History of Science*, 14, 135–154.

Hilaire-Pérez, L. (2008). Inventing in a world of guilds: The case of the silk industry in Lyon in the 18th century. In *Guilds and Innovation in Europe, 1500-1800*, Epstein, S.R. and Prak, M. (eds). Cambridge University Press, Cambridge.

Hilaire-Pérez, L. (2013). *La pièce et le geste. Artisans, marchands et savoirs techniques à Londres au XVIIIe siècle*. Albin Michel, Paris.

Hilaire-Pérez, L. (2017). Les réparations et "l'espace de la technique" au XVIIIe siècle : entre administration des pratiques et économie du produit. In *Penser la technique autrement. Mélanges en l'honneur d'Hélène Vérin*, Carvais, R., Garçon, A.-F., Grelon, A. (eds). Garnier, Paris.

Hilaire-Pérez, L. and Lanoë, C. (2011). Pour une relecture de l'histoire des métiers : les savoirs des artisans en France au XVIIIe siècle. In *La grande chevauchée. Faire de l'histoire avec Daniel Roche*, Milliot, V., Minard, P., Porret, M. (eds). Droz, Geneva.

Hilaire-Pérez, L. and Thébaud-Sorger, M. (2006). Les techniques dans l'espace public. Publicités des inventions et littérature d'usage en France et en Angleterre au XVIIIe siècle. *Revue de synthèse*, 2, 393–428.

Hilaire-Pérez, L. and Thébaud-Sorger, M. (2008). Les techniques dans la presse d'annonces au XVIIIe siècle en France et en Angleterre : réseaux d'information et logiques participatives. In *Des techniques dans la presse à la presse technique*, Bret, P., Chatzis, K., Hilaire-Pérez, L. (eds). L'Harmattan, Paris.

Hilaire-Pérez, L., Simon, F., Thébaud-Sorger, M. (eds) (2016). Inventer. In *L'Europe des sciences et des techniques XVe-XVIIIe siècles. Un dialogue des savoirs*. PUR, Rennes.

Lanoë, C. (2008). *La poudre et le fard. Une histoire des cosmétiques à la Renaissance*. Champ Vallon, Seyssel.

Lanoë, C. (2015). The artisanal economy through the prism of practical actions: The case of the Parisian glover-perfumers in the 18th century. *Journal of Innovation Economics and Management*, 3(18), 185–199.

Lanoë, C. (2021). "Mettre les gants en couleurs". La construction des savoirs des gantiers-parfumeurs, XVIe-XVIIIe siècles. *Zilsel*, 9, 219–236.

Lanoë, C. (2024). *Les ateliers de la parure. Savoirs et pratiques des artisans, France, XVIIe-XVIIIe siècle*. Champ Vallon, Cézerieu.

Le Roux, M. (2007). Un découvreur de médicaments et ses archives. Pierre Potier chercheur à l'ICSN-CNRS (1960-2005). In *Les archives de l'invention. Ecrits, objets et images de l'activité inventive des origines à nos jours*, Corcy, M.-S., Demeule-naere-Douyère, C., Hilaire-Pérez, L. (eds). CNRS, Toulouse.

Le Roux, M. (2013). Recherche académique, recherche industrielle et chimie des substances naturelles : entre l'Etat et le marché. In *Entreprises de haute technologie: État et souveraineté depuis 1945*, Fridenson, P. and Griset, P. (eds). CHEFF, Paris.

Le Roux, M. (2018). Invention collective, communautés inventives [Online] Available at: https://afhe.hypotheses.org/11263 [Accessed 24 April 2020].

Leenhardt, S. (2021). Les usages du terme "technologie" dans les discussions parlementaires sur le Plan en France (1955-1989). *Artefact*, 15, 107–126.

MacLeod, C. (2007). *Heroes of Invention. Technology, Liberalism and British Identity 1750-1914*. CUP, Cambridge.

Mercelis, J., Galvez-Behar, G., Guagnini, A. (eds) (2017). Commercializing science: Nineteenth-and-twentieth-century academic scientists as consultants, patentees, and entrepreneurs. *History and Technology*, 33(1).

Mertens, J. (2002). Technology as a science of the industrial arts: Louis-Sébastien Lenormand (1757-1837) and the popularization of technology. *History and Technology*, 18(3), 203–231.

Mialet, H. (2008). *L'entreprise créatrice. Le rôle des récits, des objets et de l'acteur dans l'invention*. Hermès-Lavoisier, Paris.

Mokyr, J. (2009). *The Enlightened Economy: An Economic History of Britain 1700-1850*. Yale University Press, New Haven.

Montègre, G. (2024). *Voyager en Europe au temps des Lumières. Les émotions de la liberté*. Tallandier, Paris.

Nègre, V. (2017). *L'art et la matière. Les artisans, les architectes et la technique entre XVIIIe et XIXe siècles*. Garnier, Paris.

Neumann, C. (2022). La mise en récit managériale des ordinateurs en France dans les années 1960. *Artefact*, 16, 111–129.

Pautet, S. (2022). Une identité matérielle en suspens : réparations, imitations et finitions des porcelaines dans l'économie de la nouveauté au XVIIIe siècle. In *Les réparations dans l'histoire. Cultures techniques et savoir-faire dans la longue durée*, Bernasconi, G., Carnino, G., Hilaire-Pérez, L. and Raveux, O. (eds). Presses des Mines, Paris.

Ratcliff, M. (2022). Formes de l'absence et production de nouveauté. In *Produire du nouveau ?* Baudry, J., Blanc, J., Hilaire-Pérez, L., Ratcliff, M., Wenger, S. (eds). Éditions du CNRS, Paris.

Roche, D. (1981). *Le peuple de Paris*. Aubier-Montaigne, Paris.

Roche, D. (1993). *La France des Lumières,* Fayard, Paris.

Roche, D. (1997). *Histoire des choses banales. Naissance de la consommation XVIIe-XIXe siècle*. Fayard, Paris.

Schlanger, J. (1983). *L'invention intellectuelle*. Fayard, Paris.

Smith, P.H. (2004). *The Body of the Artisan. Art and Experience in the Scientific Revolution*. University of Chicago Press, Chicago.

Thébaud-Sorger, M. (2009). *L'aérostation au temps des Lumières*. Presses Universitaires de Rennes, Rennes.

Thébaud-Sorger, M. (2018). Capturing the invisible: Heat, steam and gases in France and Great Britain, 1750-1800. In *Compound Histories: Materials, Governance and Production, 1760-1840*, Roberts, L. and Werrett, S. (eds). Brill, Leiden.

Thébaud-Sorger, M. (2020). Changing scale to master nature: Promoting small-scale inventions in eighteenth-century France and Britain. *Technology and Culture,* 61(4), 1076–1107.

Vérin, H. (1993). *La gloire des ingénieurs. L'intelligence technique du XVIe au XVIIIe siècle.* Albin Michel, Paris.

Vérin, H. (1998). La réduction en art et la science pratique au XVIe siècle. In *Institutions et conventions. La réflexivité de l'action économique*, Salais, R., Chatel, E., Rivaud-Danset, D. (eds). EHESS, Paris.

Vérin, H. (2002). Les machines hydrauliques dans les théâtres de machines (XIVe-XVIIe siècle). In *Archives, objets et images des constructions de l'eau du Moyen Age à l'ère industrielle*, Hilaire-Pérez, L., Massounie, D. and Serna, V. (eds). ENS Editions, Lyon.

Vérin, H. (2008). Rédiger et réduire en art : un projet de rationalisation des pratiques. In *Réduire en art : la technologie de la Renaissance aux Lumières*, Dubourg-Glatigny, P. and Vérin, H. (eds). Éditions de la Maison des sciences de l'homme, Paris.

Verna, V. (2000). Réduction du fer et innovation : à propos de quelques débats en histoire sociale des techniques. *Médiévales*, 39, 79–95.

Verna, C. (2002). *Le temps des moulines. Fer, technique et société dans les Pyrénées centrales (XIIIe-XVIe siècles).* Publications de la Sorbonne, Paris.

Woronoff, D. (1998). Le quotidien des techniques : de la répétition aux aménagements. In *Les entreprises et leurs réseaux. Hommes, capitaux, techniques et pouvoirs, XIXe-XXe siècle*, Barjot, D. and Merger, M. (eds). Presses de l'Université Paris-Sorbonne, Paris.

4

Materials and the History of Techniques

Valérie NÈGRE
IHMC, Université Paris 1 Panthéon-Sorbonne, Paris, France

4.1. Introduction

Compared with the concepts of invention, industry and technology discussed in this book, the concept of material has rarely been discussed by historians of techniques[1]. This is not to say that there has been no discussion. It is the exchange of ideas and their interdisciplinary nature that are rather recent. Over the past two decades, collective studies focusing on a single material or group of materials have proliferated. Examples include meetings and exchanges on metals, earth, marble, aluminum and "ancient and heritage materials"[2]. The thematic field of materials is

1 Special thanks to Bernadette Bensaude-Vincent and Catherine Verna for their careful review of this article.

2 "*Rencontres interdisciplinaires sur les métaux*" (2015–2023) led by Philippe Dillmann, Maxime L'Héritier and Gaspard Pagès. Available at: www.mshmondes.cnrs.fr/evenements/rencontres-interdisciplinaires-sur-les-metaux-rims. "*Échanges transdisciplinaires sur la construction en terre*" (2001–2019) led by Claire-Anne de Chazelles and Alain Klein. "*Création et Aluminium – de la découverte d'un matériau industriel à sa constitution en objet patrimonial: invention, innovation, marchés (XIXe-XXIe siècles)*" project coordinated by Patrick Fridenson; International Network for Research on Marble and Decorative Stones (NeReMa 2020). "*Matériaux anciens et patrimoniaux*" collective project, Ile-de-France research network led by Etienne Anheim, Loïc Bertrand and Isabelle Rouget, funded by the Ile-de-France regional council as part of the "*Domaines d'intérêt majeurs*" call for projects. See www.dim-map.fr.

History of Techniques,
coordinated by Guillaume CARNINO and Xavier GUCHET. © ISTE Ltd 2026.

an open interdisciplinary field with variable geometry. From one project to the next, the composition of the teams varies: while the metal specialists encourage the cross-fertilization of history, archaeology, ethnography and archaeometry, the *ancient and heritage materials* collectively aim to bring together "history, art history, archaeology, paleontology and text sciences" with "physics, chemistry, computer science, environmental sciences and biology". In the case of history, materials are used to connect the history of science with the history of techniques and to connect these same fields with the history of art or the more recent history of the environment.

The absence of an institutionalized field of study and fluctuating interdisciplinarity make any historiographical approach difficult. However, there is one movement that dominates. Materials are now seen as subjects of inquiry in their own right, *personae* capable of renewing our understanding of societies. Specialists in science and technology are unanimous in emphasizing the novelty of this approach. The forms of knowledge and practices arising from the manipulation of materials have long eluded the history of science, which focuses on theoretical knowledge, and the history of techniques, which focuses on machines, major inventions and the grand narratives of industrialization (Klein and Spary 2010, p. 3). Ultimately, few historians of techniques would have been interested in materials as such (Anonymous 1991, p. 163; Hachez-Leroy and Mioche 2012, p. 7). According to the title of the recent historiographical volume, *Voir les savoirs. Lieux, objets, gestes de la science* (Bert and Lamy 2021), the sociology of science of recent decades has examined places, objects and actions in detail, but less so materials. For anthropologist Tim Ingold, even studies of material culture have neglected materials in favor of objects – understood as matter plus form (Ingold 2007, 2012; Küchler and Were 2009). The same can be said of archaeology and art history, two disciplines that pay close attention to the materiality of objects (Conneller 2011). Technical art history lacks a theoretical framework for highlighting and interpreting the meanings of materials (Lehmann 2014; Kretzschmar 2020). Thus, contrary to what the expression suggests, the "material turn" taken by the humanities and social sciences over the last 20 years has not led to proper consideration of the subject matter and materials.

The publication of *The Social Life of Materials* (Drazin and Küchler 2015) 26 years after *The Social Life of Things* (1989), as well as the multiplication of collective volumes and monographic exhibitions devoted to stone, bronze, gold, earth, plastic or other substances, signals a new turning point. The recent volume of *Biographies of Materials* published in the *World Scientific Encyclopedia of the Development and History of Materials Science* makes this clear: renewed attention to materials is evident in many disciplines (Bensaude-Vincent 2022). Among the guiding ideas of this work, one that is constantly asserting itself is the *power* and *vitality* of materials. Materials are not passive but active and are endowed with "agency" or "affordance", depending on the author (Gallopin and Guillaume-Pey 2021). Noticeably, the criticisms of hylomorphic dualism voiced by Gilbert Simondon, Gilles Deleuze and Félix Guattari have resurfaced (Deleuze and Guattari 1980; Bennett 2010; Ingold 2012; Simondon 2017).

The transformation in the status of matter that we are witnessing in Western societies is no stranger to this movement. Digital technology profoundly reforms our relationship with physical reality (Picon 2018, p. 105). Moreover, materials are at the heart of major geopolitical and environmental issues. A new "war" is being waged over "strategic materials", a term referring to 50 rare metals that are considered indispensable to industry and to information and communication technologies (Pitron 2018). We are considering the development of "ecomaterials" and so-called "smart materials" capable of modifying their behavior according to their environment in order to meet the requirements of sustainable development. These so-called "materials by design" or "functional" materials, designed since the mid-20th century to meet specific purposes (temperature, pressure, friction, wear, insulation, electrical conductivity, etc.), blur the classical dualisms between organic and mechanical, animate and inert, as well as between science and technology (Hui 1991; Bensaude-Vincent 2004, p. 46).

In this section, we focus on interdisciplinary dynamics and show how materials studies have contributed to deepening or reconsidering three major themes in the history of techniques: innovation, interactions between science and technology, and the relationship between techniques and culture. In conclusion, we will look at how materials studies are also being used to nurture new transnational and environmental approaches.

Notably, this chapter is concerned with the concept of material, not that of matter, which is understood as an abstract and general entity. It excludes raw or elemental materials (earth, air, fire, water, etc.) and addresses the natural or artificial substances from which the artifacts around us (except food) are made.

This chapter argues that materials were not as neglected as recent historiography suggests. After all, prehistorians have chosen to delineate the great stages of civilization (Stone Age, Bronze Age, Iron Age, for example) by reference to the working of a new material, as if "a single material was capable of shaping a culture" (Bensaude-Vincent 1998, p. 20). Did André Leroi-Gourhan not – by his own admission – place matter at the heart of his investigations, stating that "it is matter that conditions all techniques, not means or forces" (Leroi-Gourhan 1971,

p. 19)? As early as the mid-1930s, did he not develop an innovative, non-anthropocentric approach, considering that objects largely work for their own evolution? He wrote that "the presence of stone gives rise to a wall, and the erection of the wall gives rise to the lever or hoist" (Leroi-Gourhan 1971, p. 27).

In these same years, Lewis Mumford (1895–1990), Henri Focillon (1881–1943), and later Gaston Bachelard (1884–1962), Cyril Stanley Smith (1903–1992) and Gilbert Simondon (1924–1989) did indeed consider materials as objects of investigation in their own right. Notably, before these works, architect and art theorist Gottfried Semper aspired to produce a general theory of the transposition of materials (*Stoffwechsel*), embracing the totality of the arts (Semper [1860] 2007).

If these reflections are not pursued further, the reason is that the task is far from easy. In his plea for a "rematerialization" of knowledge, philosopher François Dagognet (1924–2015) emphasized the extent to which the matter laid out before us immediately presents us with a challenge by asking the following: "Is it possible to leave the prison of our consciousness in order to explore it, invade it, and then master it?" (Dagognet 1985, p. 7 and 17). Before him, Gilbert Simondon had pointed out the main obstacle to observing these "acting forces". Talking about the making of clay bricks, he wrote: "In order to know the true hylomorphic relationship, it is not enough to enter the workshop and work with the craftsman: one would have to enter the mold itself, in order to follow the operation of taking form at the different levels of magnitude of physical reality" (Simondon 2017, p. 46). Entering the mold to put oneself in the place of the clay, identifying the material, dating it, then leaving it to return to the craftsman's workshop – in other words, observing it at all its scales – requires a blend of science, anthropology and history (Amouretti and Comet 1998, p. 10). It is no coincidence that those who pioneered the relationship between techniques and materials shared a taste for manual activities.

Lewis Mumford and André Leroi-Gourhan are to be taken seriously when they claim to have derived some of their hypotheses from manual experiments: radio manufacture for the former (Mumford 2016, p. 28) and axe and flint manipulation for the latter (Leroi-Gourhan 1971, p. 13[3]).

---

3 "My intuition, drawn from my own experience, provoked this new analysis. By the age of twelve, I had made my first radio" (Mumford 2018, p. 28). André Leroi-Gourhan writes: "Taking advantage, despite my theoretical training, of a rather lively taste for manual activity, I have, without any preconceived method, handled the axe, carved the flint, fired the bow and blown into the blowpipe" (Leroi-Gourhan 1971, p. 13). As far as Gilbert Simondon is concerned, see his commitment to introducing manual work into science teaching at the Lycée Des-Cartes in Tours, where he set up a technology workshop in the basement. His students built a television set.

4.2. A new understanding of innovation

To understand how the study of materials has contributed to a new understanding of innovation, we need to look back in time. The first European and North American historians of techniques, whose aim was to uncover the links between major technical changes and major economic and social evolutions, included materials in their analyses. In the models or explanatory schemes built by Lewis Mumford (1895–1990) in the United States (Mumford 2016), which were partly taken up and developed by Bertrand Gille (1920–1980) (Gille 1978) in France, materials are one of the determining components of the "complex" or "technical system". Mumford proposed a new periodization of industrial history in three overlapping phases, each determined by an energy-material "complex". Water and wood characterize the first phase (eotechnics, 1000–1750), coal and iron characterize the second phase (paleotechnics, 1750–1850) and electricity and alloys characterize the third phase (neotechnics, 1850 onward). His focus on wood led him to note that technical innovations did not follow one another but overlapped: the "culture and technique of wood" survived the age of metal and synthetic compounds (Mumford 2016, p. 98). This detailed understanding of technical change is accompanied by a pioneering approach to the relationship between materials and human culture, which we will return to later.

In comparison, Bertrand Gille's work, starting with his thesis on the iron industry in France (1943), appears to be deeply influenced by economics. The historian sets himself the task of highlighting the inherent logic of technical evolution and its links with major socioeconomic movements. For Bertrand Gille, technical progress and materials are inextricably linked: "Aviation without aluminum, the automobile without oil are as inconceivable as our current agriculture without chemical fertilizers", he writes (Gille 1979, p. 10). Among the causes of imbalance or blockage in a technical system is the scarcity of raw material. Conversely, the search for replacement products can be a source of innovation, as in the case of wood replaced by coke in England in the early 18th century or ersatz products in Hitler's Germany. One innovation leads to another. When the steel industry uses steam engines, increasingly resistant metals are needed to withstand high pressure and then overheating (Gille 1978, p. 17). In the chapter entitled "*Vers un système technique contemporain*" ("Towards a contemporary technical system"), Gille sees "new materials", in particular synthetic materials and all materials invented to meet a specific use, as a "major mutation" likely to involve "all the other techniques" (Gille 1978, p. 860).

Materials thus occupy an important place in the history told by the first historians of techniques. This remark highlights the idea that they have downplayed materials. However, the macroeconomic approach includes them in grand narratives where they serve to highlight the links between major technical advances and economic movements. This explains the primacy given to strategic industrial materials such as metals, which, for a long time, provided the bulk of weapons and tools and played a key role in industrialization, or the interest shown in materials produced by the chemical industry, such as plastics. The title given by Charles J. Singer (coeditor of one of the first general histories of techniques, *A History of Technology*, 1954–1958) to his monograph, *The Earliest Chemical Industry, an Essay in the Historical Relations of Economics and Technology Illustrated from the Alum Trade* (1948) is revealing. Alum is used to *illustrate* the relationship between technique and economics.

In France, in the 1990s, renewal did not come from a critical distancing from these early approaches, nor from a reflection on the concept of material, but from the very methods of interdisciplinary investigation that started from a real comparison of materials with the vocabulary of texts. A better understanding of how iron fits into a "technical system and economic context" and how the innovations introduced into its modes of production modify the system led researchers to change their focus of observation (Benoit 1998; Benoit and Lardin 2000). Analyses that are now local, anchored in territories, enable textual and iconographic sources to be compared with the material remains studied by archaeologists, metallographers and archaeometers (Dillmann and L'Héritier 2016[4]; Verna and Dillmann 2018). What emerges from these localized studies is a less linear history of innovation, highlighted by different types of innovation: major for some, limited for others. Emphasis is placed on the plurality of production methods, the coexistence and hybridization of old and new processes, and the diversity of rhythms in the circulation of innovations (Garçon 1998; Verna 2000). By providing data on the quality of iron, archaeometry raises new questions about the relationships among production, products and markets. In general, the materials we use link the history of production to the history of markets and consumption. In the United States, the technical historian Thomas Misa is renewing the history of steel on the basis of the concept of "user–producer

4 Archaeometry involves "information recorded by ancient objects, artifacts, or environmental archives, at various scales, most often observable through the instrumental measurement of parameters that are inaccessible to visual observation. Its methods draw on disciplines in the chemical and physical sciences, earth and life sciences, and environmental sciences" (Dillmann and L'Héritier 2016, p. 63).

interactions"[5] (Misa 1995), whereas in France, historians specializing in this same material are adopting a "new reading grid", enabling them to cross-reference the classical history of production processes and installations with that of products and markets (Hachez-Leroy 1999; Mioche and Woronoff 2006; Barjot and Bertilorenzi 2014). Plastics, which we will come back to (section 4.4), encourage the study of an innovation model that includes users and involves marketing.

Another interesting example is that of building materials. These are "detached" from the monument, so to speak, to become objects of study in their own right (Bernardi 2020, p. 7). Stone is no longer studied simply to date and document the evolution of buildings but for its own sake. Geologists, archaeologists and historians use the notion of the "*chaîne opératoire*" ("operational sequence") to highlight the dynamism of stone, from extraction to installation (Marchi and Tourneur 2004; Blary and Gély 2020[6]). Such approaches tend to link formal innovations and technical constraints. In the case of cathedrals, for example, they highlight the role of metal and lead in the "writing of the langage of the early Gothic" (Timbert 2008, p. 26).

The attention given to a wider range of materials (earth, terracotta, wood, plaster, lime, but also reused materials) considerably increases the practices and innovations that can be identified (Bernardi and L'Héritier 2018, p. 34[7]). One example is the imitation of luxury or semi-luxury materials. The increase in the number of fake materials invented from economic materials (stucco, plaster, paper, cardboard, etc.) raises the question of the role of new materials or "material innovation" in market development (Berg 2007). Far from slowing innovation, imitation appears to drive invention (Benhamou 1991; Clifford 1999; Nègre 2016). This is also what architectural and construction historians have shown about concrete. Research into the composition of Roman mortar, as well as observations of traditional adobe techniques, are at the origin of contemporary cements and concretes (Simonnet 2005; Gargiani 2013; Cellauro and Richaud 2020). Nevertheless, in the field of building materials, the shift in focus from materials considered innovative (iron, cement, reinforced concrete) (Lemoine 1991; Delhumeau 2000; Denoël et al. 2013) to traditional materials transformed by industry reveals a multiplicity of refinements and inventions with an impact on markets (Guillerme 1991b, 1995). Taking materials as a starting point means reassessing the role of innovations, which were previously

5 His work is presented in *Technology and Culture* as "The triumph of a new analytical history [of technology]", 37(3), 635–636.

6 See the works produced by Jean-Pierre Gély, Jacqueline Lorenz, and François Blary; on marble, the works of Geneviève Bresc-Bautier, Sophie Mouquin and Pascal Julien.

7 There are too many studies on wood, architectural terracotta, plaster and other materials to mention here.

seen as secondary from an economic or technical point of view, as well as taking into consideration the question of qualities and forms.

4.3. Forms of knowledge based on matter?

Another central issue that materials studies have helped revive is the complex relationship between science and technology. In 1999, Sungook Hong remarked that "no other issue in the history of science and technology has been debated as comprehensively and without consensus as the relationship between science and technology" (Hong 1999, p. 289). Indeed, since the 1930s, the question has been debated constantly in Europe and America. In France, in his foundational article setting historians the task of developing the history of techniques ("*Qu'est-ce que faire l'histoire des techniques?*"[8] in 1935), Lucien Febvre considered it essential to assess science's role in technical invention and the insertion of technical invention into the series of scientific facts. To enter into the debate on the relationship between science and technology would be to stray from our subject. Nevertheless, it is worth recalling that, in the 1960s, criticism of science and the forms of power and domination it generated were accompanied by an intense discussion of the applied-science thesis (science as the origin of technical inventions and innovations). Moreover, the emergence of materials science and engineering in the United States in the 1960s – a hybrid discipline that reversed the pure/applied schema – brought society and politics to the heart of research. Calls for a reversal of the relationship between science and invention open up a variety of avenues for research. Some aim to reveal the originality and autonomy of techniques in relation to the sciences; others aim to highlight the borrowings, overlaps and continuities between these fields. Returning to Edgard Zilsel's thesis that "modern science" was born from the merging of the rational thought of academics and humanists, on the one hand, and that of the experimental methods of "superior craftsmen", on the other hand[9] (Zilsel 1942), historians set themselves the task of observing the convergence between making and knowing (Smith et al. 2014; Chandelier et al. 2017). The notions of artisanal epistemology, vernacular knowledge and practical knowledge have been among the tools forged over the last 20 years to describe this continuum. The hypothesis behind these notions is that there are forms of knowledge (and not just know-how), or even knowledge systems (Smith 2022), derived from the interaction of the human hand with the material world. In other words, craftsmen would not only proceed by trial and error but also by systematically reflecting on their experiences, that is, experiences that produce generalizable knowledge. The

8 "What is the history of techniques?".

9 These methods are experimentation, causal interest and quantitative methods.

following question then arises: How can we identify these forms of practical knowledge, and how can we grasp how they relate to theoretical knowledge?

This is where materials prove to be excellent objects of investigation. This is because, as technical historian Francesca Bray noted, “Science is not only knowledge about matter; it is also knowledge that comes from matter” (Bray 2008, p. 323). The material thinking of craftsmen and the analytical thinking of scientists, points out metal archaeologist Maikel Kuijpers, lead to different choices of categorization and metaphor, “but both are equally dependent on the physical properties of matter that anchor the knowledge produced. These two epistemologies must therefore be compatible with one another. Science and craft do not describe different material realities; they simply describe reality differently” (Kuijpers 2019, p. 609). To grasp this knowledge that comes from matter, historians do not hesitate to step outside historical knowledge to call on methods of reconstruction or restitution that have long been practiced by antiquarians, architects and archaeologists (Dupre et al. 2020).

In particular, historians of science, Ursula Klein and Emma Spary, emphasize the value of “useful material substances” (Klein and Spary 2010, p. 1). By this expression, the two historians mean materials that are bought and sold, consumed and subjected to markets. Compared with the “classical” objects of the history of science, such as invisible substances (atoms, magnetic forces) – which can only be detected and manipulated through complex devices – or scientific instruments (air pumps, electrical machines, etc.) handled by scientists and their assistants, useful material substances produced and exchanged by many individuals (craftsmen, traders, amateurs, users) document common practices and knowledge that are ordinary and learned.

The study of these substances shows that certain craft experiments are at the origin of concepts that have long been considered to be the result of modern science. According to Pamela Smith, one of the most widespread and enduring theories on matter and its transformation in the field of metallurgy – the sulfur–mercury theory – arose from the manufacture of materials. According to Marco Beretta, master glassmakers’ attempts to imitate precious and natural stones laid the foundations for the transmutation theory of alchemy (Beretta 2009; Smith 2010). As early as 1960, Cyril Stanley Smith, director of the Institute for the Study of Metals and president of the Society for the History of Technology, urged historians of science and technology to reintegrate the sensitive experiences of craftsmen into the general narrative of knowledge production (Smith 1960).

Another advantage of ordinary materials over the hidden entities observed by scientists is that they provide visibility to the mental structures underlying scientific and technical practices. Clay, metal, wood and pigments not only have physical properties (objective, scientifically measurable) but also possess cultural meanings and values that vary considerably according to historical and geographical context. In this way, they provide access to different "worldviews" that underpin manufacturing practices (Smith 2010). While Gaston Bachelard sought to reveal a general ontology of material imagination based on the reveries of poets, the close-up scale of observation here offers the possibility of grasping the specific and multiple ontologies of operators. The materials thus encourage us to include culture in our reinterpretation of the relationship between science and technology. Ultimately, the questions presented by historians of science and technology coincide with those presented by historians of culture (Lehmann et al. 2013; Anderson et al. 2014).

4.4. Reintroducing techniques into culture

In the second preface to *Technics and Civilization*, written in 1963, Lewis Mumford felt that his 1934 book "heralded a change of attitude" among scholars, who were beginning to consider "the history of technics as an element in human culture and, to a lesser degree, toward evaluating its social and cultural results" (Mumford 2010, p. xvii). With respect to materials, he emphasized the importance of wood and glass culture and techniques in the modern and contemporary eras. However, few historians have taken up these ideas. The false opposition between culture and technique denounced by Gilbert Simondon in the 1950s has unknowingly influenced historical approaches. The philosopher stated that technical objects have long been repressed "in the unstructured world of things that have no meaning, but only a use, a useful function" (Simondon 2012, p. 10). While the publications of the *Technology and Culture* (1959) journals in the United States, *Culture Technique* (1979–1993) and *Techniques and Culture* (1983) in France signaled a change in attitude, at the dawn of the 21st century, Francesca Bray was still calling for techniques to be treated as a form of cultural expression (Bray 1998, p. 33). According to *The Business History* (Hounshell 1985), the avenues opened up by Lewis Mumford with respect to materials, particularly wood, were not explored until the collective volume edited by Brooke Hindle (1918–2001). It then became clear that wood, owing to its abundance and low cost, dominated the material culture of the first century and half of the "American era" (Brooke 1982).

Interestingly, some materials themselves invite us to reintroduce techniques into culture. This is particularly true of materials such as plastic – "everything swallowed up in its use" (Barthes 1991) – which many eyes have turned to since the 1950s.

This new, multifunctional material, which is largely replacing many traditional substances (wood, ceramics, metal, ivory, horn, etc.) and finding applications in a large number of everyday objects, prompted us to look at the cultural factors that encourage its spread. Thus, a few years after the publication of *The Plastic Age, 1917–1930* (1971) by the historian of American civilization Robert Sklar, whose title suggested the power of plastic to shape American civilization, the technical historian Robert Douglas Friedel added *Pioneer Plastic* (1983) to his classic technoeconomic survey, an analysis of how artificial plastics are seen and used by consumers (Friedel 1983). Friedel shows how a celluloid's image affects its marketing. Its "plasticity" – the fact that it can take on a variety of shapes and colors – initially led to its association with the idea of a cheap, inferior material. Friedel thus moved away from the early histories of plastics centered on industrial production and innovation. In addition to the traditional questions posed by historians of techniques, he also adds those related to the role of usage and cultural determination in product development and innovation. His point of view is similar to that of cultural historian Jeffrey Meikle, who, in *American Plastic. A Cultural History* (1995), looks at the process by which plastic comes to symbolize "modernity" and the "aura of the news" (Meikle 1995, p. 57 and 3). Nylon and aluminum also raise similar questions (Handley 1999; Friedenson and Hachez-Leroy 2017). Éric Schatzberg's investigation into aluminum leads him to conclude that symbolic values clearly influence technical innovations (Schatzberg 2003).

Certain types of material call for ambitious research that combines different methodological approaches. Coal offers Jean-François Mathis the opportunity to grasp ordinary life and, through it, the workings of an "energy system, underpinned by political, economic, social and cultural structures and choices" (Mathis 2021, p. 18). Using silver objects produced in English society in the second half of the 18th century as a starting point, Hélène Clifford delves into what continues in manufacturers' workshops, retailers' stores and buyers' homes; in other words, she links the history of manufacture to that of consumption and use (Clifford 2004, p. 2). In the catalog for the exposition of *Gold, Power and Allure* (2012), the historian notes that gold calls for the connection of objects usually thought of separately by geologists, metallurgists, archaeologists, economists and historians (geological specimens, prehistoric jewelry, Enlightenment-era crockery, etc.) (Clifford 2012). Similarly, metal leads Cyril Stanley Smith, a historian of science and technology, to observe objects held in art museums, thereby reassessing the role of artists in the evolution of techniques. He states that "technology is more closely related to art than to science – not only materially, because art must somehow involve the selection and manipulation of matter, but conceptually as well, because the technologist, like the artist, must work with many unanalyzable complexities". For Smith, necessity is not the mother of invention but only of refinement: "Discovery requires aesthetically motivated curiosity, not logic" (Smith 1976, p. 144). His investigation brought him closer to the reflections of art historian Henri Focillon. In the third chapter of *La vie des formes* (1943), meaningfully entitled "*Les formes dans la matière*" ("Forms in matter"), Focillon suggests eliminating the "old antinomies, mind-matter, matter-form", to stop "considering form, matter, tool and hand in isolation" and to place ourselves "at the meeting point, at the geometric place of their activity" (Focillon 1955, p. 62). He writes that form "does not act as a superior principle shaping a passive mass, for matter can be said to impose its own shape on form" (Focillon 1955, pp. 51–52). The historian's role is to consider what he calls the "formal vocation" of matter. The idea is similar to the concept of "natural form" developed by Gilbert Simondon a few years later. For the great specialist of techniques, matter itself conceals the coherence of form before it takes shape. Knowing how to use a tool is not just a matter of an action but of recognizing the "implicit form" or "natural form" of matter (Simondon 2012, p. 46). Simondon proposed the idea that the intrinsic optimization of a technical object consists of thinking about the material, form and function together. He thus invites us to look at alloys as the most "concrete" possible adhesion of matter to form and vice versa (Simondon 2014). Few historians have followed the paths laid out by these authors. Nevertheless, we should mention the pioneering study by art historian and Victoria and Albert Museum curator Mikael Baxandall, *The Limewood Sculptors of Renaissance Germany* (Baxandall 1980), which takes the analysis of hidden forms in wood as a starting point for examining the social and political world in which Renaissance sculptors evolved. Following Baxandall's footsteps, and for the past three decades, historians of sculpture, architecture and construction have been trying to bring together a history of production and a history of reception[10].

4.5. Conclusion

Historically, the process of abstracting materials from objects and technical actions is not obvious. In an article entitled "*Comment analyser et décrire les actes techniques?*"[11] (1993), François Sigaut challenged Leroi-Gourhan's postulate: "Leroi-Gourhan 'posits in principle' that 'it is matter that conditions all technique, and not means or forces'. We fail to see the basis for this assertion, which, rather than a principle, is a petition of principle; for it is quite obvious that all the means of a technique condition that technique" (Sigaut 1993, p. 6). For Sigaut, following the

10 In regard to architecture and construction, there are too many works to mention. For sculpture, see the special issue of *Sculpture Journal* devoted to marble (2021).

11 How can technical actions be analyzed and described?

footsteps of Marcel Mauss, techniques are first and foremost actions, in other words, movements of the body and limbs, applied "to a material in order to modify it according to a more or less precise program" (Sigaut 1994, p. 19; 1999, p. 521).

However, focus *first* on matter – in other words, isolating materials from objects and actions – has repercussions on the way we think about techniques and their history. *Starting* with materials, *following* the itinerary of substances with singular characteristics, has decisive methodological repercussions. The operation implies a close scale of observation and an analysis proceeding from the specific to the general. Far from leading to a specialization of history, this approach encourages dialogue between disciplines, as well as a reflexive approach to the sources and fields (science, technique, culture) that have been used to divide the different fields of knowledge.

It is hard to deny that, for a long time, the history of science and technology has tended to neglect materials. However, this fact should not overshadow the richness of earlier reflections, which should be better considered. We have limited ourselves to showing how interdisciplinary work on specific subjects has helped to renew the themes of innovation, the relationship between science and technology, and the relationship between techniques and culture. Other choices were possible. For almost two decades, the history of materials has followed the transformations of the social sciences at a greater or lesser distance. Materials are proving to be excellent conductors for illuminating new transnational and environmental approaches, for example. Focusing on gunpowder, steel, uranium, gems, glass, indigo or other materials traded on a global scale enables collectives to revisit the history of colonial and imperial spaces (Buchanan 2006; Evans and Rydén 2007; Hecht 2012; Bycroft and Dupré 2019) or that of the circulation of knowledge and practices (Curtis 2008; Prakash 2012; Smith 2019). Aluminum, mercury and synthetics, to name but a few examples, also provide plenty of material to rewrite the history of the environment (Le Roux 2003; Robins 2011; Blanc 2020). In the long term, beyond human use, such materials lead to waste and pollution becoming historical objects.

4.6. References

Amouretti, M.-C. and Comet, G. (1998). Artisanat et matériaux. La place des matériaux dans l'histoire des techniques. *Cahiers d'histoire des techniques*, 4, 9–14.

Anderson, C., Dunlop, A., Smith, P. (eds) (2014). *The Matter of Art: Materials, Technologies, Meanings 1200-1700*. Manchester University Press, Manchester.

Anonymous (1991). Editorial. *History and Technology*, 7(3–4), 163–164. doi: 10.1080/07341519108581774.

Barjot, D. and Bertilorenzi, M. (eds) (2014). *Aluminium du métal de luxe au métal de masse (xixe-xxie siècle)*. PUF, Paris.

Barthes, R. (ed.) (1991). Plastic [Online]. Available at: https://soundenvironments.wordpress.com/wp-content/uploads/2011/11/roland-barthes-mythologies.pdf.

Baxandall, M. (1980). *The Limewood Sculptors of Renaissance Germany*. Yale University Press, New Haven.

Benett, J. (2010). *Vibrant Matter. A Plotical Ecology of Things*. Duke University Press, Durham.

Benhamou, R. (1991). Imitation in the decorative arts of the eighteenth century. *Journal of Design History*, 4(1), 1–13.

Benoit, P. (1998). L'innovation technique et ses marqueurs. Au four et au moulin : innovation et conjoncture. *Actes des congrès de la Société d'archéologie médiévale*, 6, 293–301.

Benoit, P. and Lardin, P. (2000). Les paris de l'innovation. *Médiévales*, 39, 5–13.

Bensaude-Vincent, B. (1998). *Eloge du mixte. Matériaux nouveaux et philosophie ancienne*. Hachette, Paris.

Bensaude-Vincent, B. (2004). *Se libérer de la matière ? Fantasmes autour des nouvelles technologies*. INRA Éditions, Paris.

Bensaude-Vincent, B. (ed.) (2022). *Between Nature and Society. Biographies of Materials*. World Scientific Publishing, London.

Beretta, M. (2009). *The Alchemy of Glass: Counterfeit, Imitation, and Transmutation in Ancient Glassmaking*. Science History Publications, Sagamore Beach.

Berg, M. (2007). *Luxury and Pleasure in Eighteenth-Century Britain*. Oxford University Press, Oxford.

Bernardi, P. (2020). Préface. In *Pierres de construction. De la carrière au bâtiment*, Blary, F. and Gély, J.-P. (eds). CTHS, Paris.

Bernardi, P. and L'Héritier, M. (eds) (2018). Introduction. Recyclage et remploi : la seconde vie des matériaux de construction. *Aedificare*, 4, 23–36.

Bert, J.-F. and Lamy, J. (2021). *Voir les savoirs. Lieux, objets, gestes de la science*. Anamosa, Paris.

Blanc, P.D. (2020). *Fake Silk: The Lethal History of Viscose Rayon*. Yale University Press, New Haven.

Blary, F. and Gély, J.-P. (2020). *Pierres de construction. De la carrière au bâtiment*. CTHS, Paris.

Bray, F. (1998). Technics and civilization in late imperial China: An essay in the cultural history of technology. *Osiris*, 13, 11–33.

Bray, F. (2008). Science, technique, technology: Passages between matter and knowledge in imperial Chinese agriculture. *British Society for the History of Science*, 41(3), 319–344.

Brooke, H. (ed.) (1982). *Material Culture of the Wooden Age*. Sleepy Hollow Press, Tarrytown, NY.

Buchanan, B.J. (ed.) (2006). *Gunpowder, Explosives and the State: A Technological History*. Ashgate Publishing, Farnham.

Bycroft, M. and Dupré, S. (2019). *Gems in the Early Modern World: Materials, Knowledge and Global Trade, 1450–1800*. Palgrave Macmillan, Basingstoke.

Cellauro, L. and Richaud, G. (2020). *Palladio and Concrete. Archaeology, Innovation, Legacy*. L'Erma di Bretschneider, Rome.

Chandelier, J., Verna, C., Weill-Parot, N. (eds) (2017). *Science et technique au Moyen Âge*. Presses Universitaires de Vincennes, Paris.

Clifford, H. (1999). Concepts of invention, identity and imitation in the London and provincial metal-working trades, 1750-1800. *Journal of Design History*, 12(3), 241–255.

Clifford, H. (2004). *Silver in London: The Parker and Wakelin Partnership 1760-1776*. Yale University Press, New Haven.

Clifford, H. (ed.) (2012). *Gold: Power and Allure*. Worshipful Company of Goldsmiths, London.

Conneller, C. (2011). *An Archaeology of Materials: Substantial Transformations in Early Prehistoric Europe*. Routledge, London.

Curtis, E.B. (2008). *Glass Exchange between Europe and China, 1550-1800: Diplomatic, Mercantile, and Technological Interactions*. Ashgate, Burlington.

Dagognet, F. (1985). *Rematérialiser*. Vrin, Paris.

Deleuze, G. and Guattari, F. (1980). *Mille plateaux*. Minuit, Paris.

Delhumeau, G. (2000). *L'invention du béton armé : Hennebique 1892-1910*. Norma, Paris.

Denoël, J.-F., Espion, B., Hellebois, A., Provost, M. (2013). *Histoire de Béton Armé : patrimoine, durabilité et innovations*. Febelcem et Fabri, Brussels.

Dillmann, P. and L'Héritier, M. (2016). Archéométrie et histoire des techniques : les procédés direct et indirect en sidérurgie (XIVe-XVIIe siècle). *Artefact*, 4, 63.

Drazin, A. and Küchler, S. (2015). *The Social Life of Materials: Studies in Materials and Society*. Routledge, London.

Dupre, S., Harris, A., Kursell, J. (2020). *Reconstruction, Replication and Re-enactment in the Humanities and Social Sciences*. Amsterdam University Press, Amsterdam.

Evans, C. and Rydén, G. (2007). *Baltic Iron in the Atlantic World in the Eighteenth Century*. Brill, Leiden.

Febvre, L. (1935). Réflexion sur l'histoire des techniques. *Annales d'histoire économique et sociale*, 7(36), 531–535.

Focillon, H. (1955). *La vie des formes*. PUF, Paris.

Frasca-Rath, A., and Kretzschmar, M. (eds) (2021). Special issue: Marble. *Sculpture Journal*, 30(2), 117–121.

Fridenson, P. and Hachez-Leroy, F. (eds) (2017). *L'aluminium, matière à création, XIXe-XXIe siècles*. PUR, Rennes.

Friedel, R.D. (1983). *Pioneer Plastic: The Making and Selling of Celluloid*. University of Wisconsin Press, Madison.

Gallopin, T. and Guillaume-Pey, G. (eds) (2021). *Ce que veulent les pierres*. Presses Universitaires de Liège, Liège.

Garçon, A.-F. (1998). *Mine et métal, 1780-1880. Les non-ferreux et l'industrialisation*. PUR, Rennes.

Gargiani, R. (2013). *Concrete, from Archaeology to Invention 1700-1769*. EPFL, Lausanne.

Gille, B. (ed.) (1978). *Histoire des techniques*. Gallimard, Paris.

Gille, B. (1979). La notion de "système technique" (essai d'épistémologie technique). *Technique et culture*, 1, 8–18.

Guillerme, A. (1991a). Editorial. *History and Technology*. 7(3–4), 163–164. doi: 10.1080/07341519108581774.

Guillerme, A. (1991b). Techniques et matériaux de construction, XIXe-XXe siècle. Une introduction. *History and Technology*, 3(4), 165–178.

Guillerme, A. (1995). *Bâtir la ville. Révolutions industrielles dans les matériaux de construction. France-Grande-Bretagne (1760-1840)*. Champ Vallon, Paris.

Hachez-Leroy, F. (1999). *L'Aluminium français. L'invention d'un marché*. CNRS, Paris.

Hachez-Leroy, F. and Mioche, P. (2012). Le matériau : un nouvel objet pour l'historien ? Ou comment saisir le processus de création continue d'un matériau sur le temps long. *Cahiers d'histoire de l'aluminium*, 49(2), 7–19.

Handley, S. (1999). *Nylon: The Story of a Fashion Revolution*. Johns Hopkins University Press, Baltimore.

Hecht, G. (2012). *Being Nuclear: Africans and the Global Uranium Trade*. The MIT Press, London.

Hong, S. (1999). Historiographical layers in the relationship between science and technology. *History and Technology*, 15, 289–311.

Hounshell, D.A. (1985). Reviewed work. "Material Culture of the Wooden Age" by Brooke Hindle. *The Business History*, 59(3), 493–495.

Hui, J. (1991). Les matériaux, une histoire vieille comme l'homme. *Culture technique*, 23, 102–120.

Ingold, T. (2007). Materials against materiality. *Archeological Dialogues*, 14, 1–16.

Ingold, T. (2012). Toward an ecology of materials. *Annual Review of Anthropology*, 41, 427–442.

Klein, U. and Spary, E. (eds) (2010). Introduction. Why materials? In *Materials and Expertise in Early Modern Europe. Between Market and Laboratory*, Klein, U. and Spary, E.C. (eds). University of Chicago Press, Chicago.

Kretzschmar, M. (2020). Inhalt – form – material. Kunsthistorische Materialforschung als Diskursfeld. In *Kultur und Technik. Interdisziplinäre Perspektiven*, Jarchow, M. and Orth, D. (eds). Wachholtz, Kiel.

Küchler, S. and Were, G. (2009). Empathie avec la matière. *Techniques and Culture*, 52(53), 190–211.

Kuijpers, M. (2019). Materials and skills in the history of knowledge: An archaeological perspective from the "Non-Asian" field. *Technology and Culture*, 60(2), 604–615.

Le Roux, M. (2003). L'environnement et les industries chimiques : le cas de l'aluminium. In *Les sources de l'histoire de l'environnement*, Corvol, A. (ed.). L'Harmattan, Paris.

Lehmann, A.-S. (2014). The matter of the medium. Some tools for an art theoretical interpretation of materials. In *The Matter of Art: Materials, Technologies, Meanings 1200-1700*, Anderson, C., Dunlop, A., Smith, P.H. (eds). Manchester University Press, Manchester.

Lehmann, A.-S., Scholten, F., Chapman, H.P. (eds) (2013). *Meaning in Materials, 1400-1800*. Brill, Boston.

Lemoine, B. (1991). Fer et béton en France (1850-1920). *History and Technology*, 3(4), 1267–278.

Leroi-Gourhan, A. (1971). *L'Homme et la matière*. Albin Michel, Paris.

Marchi, C. and Tourneur, F. (eds) (2004). *Pouvoir(s) de marbres*. Commission royale des Monuments, Liège.

Mathis, J.-F. (2021). *La civilisation du charbon (1830-1940)*. Vendémiaire, Paris.

Meikle, J.L. (1995). *American Plastic: A Cultural History*. Rutgers University Press, New Brunswick.

Mioche, P. and Wornoff, D. (eds) (2006). *L'acier en France : produits et marchés, de la fin du XVIIIe siècles à nos jours*. Éditions Universitaires de Dijon, Dijon.

Misa, T. (1995). *A Nation of Steel: The Making of Modern America, 1865-1925*. John Hopkins University Press, Baltimore.

Mumford, L. (2010). *Technics and Civilization*. University of Chicago Press, Chicago.

Nègre, V. (2016). *L'Art et la matière. Les artisans, les architectes et la technique (1770-1830)*. Classiques Garnier, Paris.

Picon, A. (2018). *La matérialité de l'architecture*. Parenthèses, Marseille.

Pitron, G. (2018). *La guerre des métaux rares. La face cachée de la transition énergétique et numérique*. Les Liens qui Libèrent, Paris.

Prakash, K. (2012). *Indigo Plantations and Science in Colonial India*. Cambridge University Press, New York.

Robins, N.A. (2011). *Mercury, Mining and Empire: The Human and Ecological Cost of Colonial Silver Mining in the Andes*. India University Press, Bloomington.

Schatzberg, E. (2003). Symbolic culture and technological change: The cultural history of aluminum as an industrial material. *Enterprise and Society*, 4(2), 226–271.

Semper, G. (2007). *Du Style et de l'architecture. Ecrits, 1834-1869*. Parenthèses, Marseille. [Original published 1860].

Sigaut, F. (1993). How can we analyse and describe technical actions? In *The Use of Tools by Human and Non-Human Primates*, Berthelet, A. and Chavaillon, J. (eds). Clarendon Press, Oxford.

Sigaut, F. (1994). "Technology" typescript of the French version. In *Companion Encyclopedia of Anthropology*, Ingold, T. (ed.). Routledge, London.

Sigaut, F. (1999). Revenir à une géographie des techniques. In *Dans le sillage des techniques. Hommage à Robert Cresswell*, Jamard, J.-L., Montigny, A., Picon F.-R. (eds). L'Harmattan, Paris.

Simmonet, C. (2005). *Le béton, histoire d'un matériau. Économie, technique, architecture*. Parenthèse, Paris.

Simondon, G. (2012). *Du mode d'existence des objets techniques*. Aubier, Paris.

Simondon, G. (ed.) (2014). Anthropo-technologie. In *Sur la technique*. PUF, Paris.

Simondon, G. (2017). *L'individuation à la lumière des notions de forme et d'information*. Jérôme Millon, Paris.

Smith, C.S. (1960). *A History of Mettalography: The Development of Ideas on the Structure of Metals Before 1890*. University of Chicago Press, Chicago.

Smith, C.S. (1976). On art, invention, and technology. *Technology Review*, 10, 144–147.

Smith, P.H. (2010). Vermillion, mercury, blood, and lizards: Matter and meaning in metalworking. In *Materials and Expertise in Early Modern Europe*, Klein, U. and Spary, E.C. (eds). University of Chicago Press, Chicago.

Smith, P.H. (ed.) (2019). *Entangled Itineraries: Materials, Practices, and Knowledges across Eurasia*. University of Pittsburgh Press, Pittsburgh.

Smith, P.H., Meyers, A.R., Cook H.J. (eds) (2014). *Ways of Making and Knowing: The Material Culture of Empirical Knowledge*. University of Michigan Press, Ann Arbor.

Timbert, A. (ed.) (2008). Introduction. In *L'homme et la matière. L'emploi du plomb et du fer dans l'architecture gothique*. Picard, Paris.

Verna, C. (2000). Réduction du fer et innovation : à propos de quelques débats en histoire sociale des techniques. *Médiévales*, 39, 79–95.

Verna, C. and Dillmann, P. (eds) (2018). Histoire des techniques médiévales et culture matérielle. In *La culture matérielle : un objet en question. Anthropologie, archéologie et histoire*. Presses universitaires de Caen/Brépols, Caen.

Zilsel, E. (1942). The social roots of science. *The American Journal of Sociology*, c47, 544–562.

5

Concepts of the History of Techniques: Technology

Eric SCHATZBERG
Ivan Allen College of Liberal Arts, Georgia Institute of Technology, Atlanta, USA

5.1. Introduction: an odd concept

Technology is an odd concept. It is at the center of discourse about late modernity, yet scholars rarely engage with it as deeply as other core concepts, such as science, democracy or nationalism. There are, of course, exceptions to this relative neglect, but these tend to be isolated by language and discipline. Even between European languages, translators often have trouble finding suitable equivalents. Many non-European languages simply import the English term as a loanword, as in the Hebrew טֶכנוֹלוֹגְיָה (tekhnologia) or Japanese テクノロジー (Tekunorojī), although native terms also persist alongside the loanwords, as in the Japanese 技術 (Jijitsu).

Concepts acquire meaning in part through naming, that is, through the specific words that convey their meaning. These words connect and distinguish a concept from related concepts[1]. For technology, three key related terms are industry, invention and materials, each of which is examined in a chapter in this book. Additional terms include science ("science and technology"), innovation ("technological innovation"), engineering, techniques, production, artifacts, mechanical arts and so on. The meanings of these terms often overlap, with

1 There is a vast literature on the history of concepts, for example: Burke, M. and Richter, M. (eds) (2012). *Why Concepts Matter: Translating Social and Political Thought*. Leiden, Brill.

boundaries that shift between languages and discipline. As a category, technology holds together through sometimes vague family resemblances, although the core of the concept centers around human actions to reshape the material world[2].

The fuzziness of this concept has long perplexed historians of technology. Because of terminological differences, each national tradition has grappled with the concept in its own language, although almost always in dialogue with other linguistic traditions. One of the largest divisions appeared in Europe between English usage and Continental languages: the differences between the cognates of the terms *technique* and *technology*. This distinction is deeply implicated in the history of the term and in continued confusion over its meaning.

The oddity of the English term *technology* is clear from its etymology. Although the word's origins remain somewhat obscure, its roots are in the Greek terms *techne* and *logos*, both key concepts in the Greek philosophical tradition. A large amount of literature exists on both terms. To simplify, *logos* refers to reasoned discourse, whereas *techne* is commonly translated as "art", "craft", or "skill". Words with the suffix "-logy" usually denote a field of study or branch of knowledge. However, in present-day English usage, *technology* refers more to things than to ideas or material practices rather than to a scholarly discipline. In fact, there has never been an established scholarly discipline designated by the English term *technology*.

5.2. History of technology or history of techniques?

We can trace the confused history of technology as a concept by examining the history of the technology itself. When this field emerged in Continental Europe in the first half of the 20th century, it was based not on cognates of *technology* but rather on *technique*. In Germany, the field grew out of professional engineering, specifically the national engineering society Verein Deutscher Ingenieure (VDI), as by the engineer Conrad Matschoss. In 1909, Matschoss founded one of the first scholarly periodicals dedicated to the history of technology, the annual volume *Beiträge zur Geschichte der Technik und Industrie*, published under the auspices of the VDI. In his foreword to the first volume, Matschoss made clear his goal for this publication. *Beiträge* aimed to increase the cultural status of

2 A small sample of work on these related concepts includes Schauz, D. (2015). Wissenschaftsgeschichte und das Revival der Begriffsgeschichte. *NTM Zeitschrift für Geschichte der Wissenschaften, Technik und Medizin,* 23(1–2), 53–63; Godin, B. (2017). *Narratives of Innovation: The Invention of Models, 1920-1980.* MIT Press, Cambridge; Bud, R. (2014). "Applied science" in nineteenth-century Britain: Public discourse and the creation of meaning, 1817-1876. *History and Technology,* 30(1/2), 3–36.

Technik, which Matschoss conceived largely in terms of the "masterpieces of the great engineers" (Matschoss 1909). Specifically, *Technik* for Matschoss was very closely linked with engineering, even though the scope of topics included under *Technik* was broader than the work of professional engineers. In fact, German lacks a close equivalent to the English term *engineering*, in the sense of the activity of engineers, leaving *Technik* as the default term. The German term *Technologie*, in contrast, made no appearance in Matschoss's forward.

It is no surprise that Matschoss encouraged a technical history of technology – history for engineers, not historians. Most of the articles in the first volume focused on recent topics, with the exception of two articles about Hero of Alexandria. These articles included histories of machinery for mining and metallurgy, biographies and autobiographies of engineers, accounts of steam engines, steam pressure vessels and woodworking machinery, as well as an account of the early years of the electrical manufacturer AEG.

In France, a major contribution to the history of technology emerged from the 1930s Annales school. This contribution appeared in a special issue of the *Annales d'histoire économique et sociale* published in 1935, as Pamela Long described in detail (Long 2005). This journal, founded by the innovative social historians Marc Bloch and Lucien Febvre, promoted a new type of history that went beyond the discipline's traditional focus on war, politics and great men. The special issue was *Les techniques, l'histoire et la vie*, with a remarkable introduction by Febvre titled "Reflexions sur l'histoire des techniques" (Febvre 1935). As with Mattschoss's *Beiträge*, the French term *technologie* was completely absent from the text of the articles.

Febvre's introduction was a remarkable statement. Like Mattschoss, Febvre began by decrying the lack of historical research on *les techniques*. He proposed three lines of research to help rectify this deficit. The first was the need for a "histoire technique de la technique", a technical history of technology. This was a job for technicians ("techniciens"), despite their lack of historical sensibilities. However, broad-minded technicians were capable of acting "like clever historians [...] following *les techniques* in their travels across time as well as across space". The second line of research focused on the relationship between theory and practice, or more specifically, the relationship between science and *technique*. Febvre argued that technical progress can depend on chance, a purely technical sequence of cause and effect, or the progress of science as a whole. Thus, technical invention can indeed depend on science. However, the science–technology relationship can also work in another direction: scientific facts can themselves rely on technical invention.

"Is that all?", asked Febvre. Certainly not. *Technique* cannot be considered apart from the "other human activities" that frame it. These activities include religion, art, politics and needs of the era in the broadest sense, not just immediate needs. Finally, the relationship between the general history and the history of *techniques* is bidirectional, such that "each epoch has its own *technique*, and this *technique* has the style of the epoch".

Febvre's program provided a sophisticated agenda for the new field of history of technology. No comparable program exists in the English-speaking world. The closest was Lewis Mumford's *Technics and Civilization*, published in 1934, which also argued for examining technology within its broad human context (Mumford 1934). However, there is no history of technology in the United States. Although Mumford did use the term "technology", he used *technics* as his core concept alongside *the machine*. *Technics*, as used by Mumford, most likely served as a translation of the German *Technik* (Schatzberg 2018, pp. 145–151).

However, when the history of technology emerged as an academic field in the United States in the late 1950s, *technology* rather than *technique* or *technics* was the key term, as indicated by the name of the principal organization: the Society for the History of Technology. The key push to formalize the field came from a somewhat odd place: historians and other humanities scholars who taught in America's rapidly growing engineering schools. At the time, American engineers had little use for the term *technology*, as *engineering* already provided a unifying concept for their diverse field. When, in the post-Sputnik fervor to strengthen American science, politicians proposed a cabinet-level Department of Science and Technology, engineering leaders complained about the name. *Technology*, they insisted, was not an acceptable alternative for *engineering* (Schatzberg 2018, p. 207).

Despite *technology*'s relative lack of currency in American English, the founders of the Society for the History of Technology embraced *technology* as their core term, both for the name of the Society and its journal, *Technology and Culture*. The driving force behind the new society was a professor of European history named Melvin Kranzberg, who then taught at the Case Institute of Technology. Kranzberg received his Ph.D. in modern French history from Harvard in 1942. Before joining the faculty at Case, he taught at another engineering university, the Stevens Institute of Technology. Even before moving to Case in 1952, Kranzberg worked on a project sponsored by the American Society for Engineering Education about how to improve the teaching of non-technical courses for engineering students (Schatzberg 2018, pp. 207–209).

When Kranzberg first raised the issue of creating an academic organization dedicated to the history of technology, he assembled an advisory committee that consisted of an odd mix of broadminded engineers and technically-minded historians. The committee's members knew little about the existing European scholarship on the history of *technique*, with two exceptions. The first was Lynn White, a medieval historian who served as president of Mills College in California until 1958. White's publications show that he was familiar with much of the Continental scholarship on the history of medieval *techniques* (White 1940). However, his duties as a college president prevented early discussions about the founding of SHOT and its journal. The second was Lewis Mumford. Even though Mumford had not kept up with the Continental scholarship on *technique* since finishing *Technics and Civilization* in 1934, he did have reservations about using the term *technology* in the name of the new journal. When Kranzberg proposed the name *Technology and Culture*, Mumford instead suggested *Technics and Culture*. As he explained in a letter to Kranzberg:

> there would be a great gain in differentiating between Technics and Technology, as the Germans have done. Technics is an accurate and useful term for denoting the whole field; and though technology has been used for this purpose in English since 1859, it would be better reserved for the systematic investigation of methods and processes. I realize that this usage is not yet general; but it is time that a start was made. (Schatzberg 2018, p. 150)

Mumford's private comment represents one of the few times that American scholars discussed the Continental distinction between *technique* and *technology*. However, the time had long passed for translating the Continental meanings of *Technik* using its English cognates, whether *technics* or *technique*. Mumford himself dropped the ball in 1934 in *Technics and Civilization*, where he used the terms *technics* and *technology* more or less interchangeably. Instead, *technology* would remain the core term for both the Society for the History of Technology and its journal, *Technology and Culture*.

As might be expected, the articles in the first few volumes of *Technology and Culture* devoted significant attention to defining *technology*. One commonly cited definition came from the opening essay of the first volume of Charles Singer's pioneering encyclopedia, *A History of Technology*, published in 1954. In this essay, the editors explicitly disavowed any lengthy analysis of the meaning of *technology*. They did briefly acknowledge that the term had a history, noting its initial definition as "systematic discourse about the (useful) arts". They claimed that technology had then come to be seen as "almost identical with 'applied science'". However, for their purposes, *technology* was defined "as covering the field of how things are commonly done or made," and by extension "what things are done or made" (Singer et al. 1954, p. viii). Such a definition was so broad as to be almost meaningless, as a number of early contributors to *Technology and Culture* noted. Kranzberg, for example, criticized Singer's definition in the opening essay of this new journal, noting that this definition treated technology as if it were "insulated from the rest of society". However, Kranzberg (1959, pp. 8–9) did not provide an alternative.

When other authors addressed the meaning of *technology* in the early years of *Technology and Culture*, they embraced a wide range of opinions about what it did or should mean. Some authors restricted the definition of *technology* to material artifacts, especially tools and machinery, whereas others argued for including immaterial "tools" such as language. Many definitions emphasize technology as a process with varying degrees of human agency, sometimes including skills and crafts. Other definitions described *technology* as a field of study, a type of scientific knowledge or the application of scientific knowledge. Some authors understood technology as roughly coterminous with engineering, whereas others included agriculture and even medicine. A few viewed *technology* as covering forms of organization or even as the dominant cultural system of our age (Schatzberg 2018, pp. 210–211). Such disunity was rather remarkable for a newly created academic field, which seemed to lack a basis for shared discourse.

In the 1970s and 1980s, when Europeans began taking note of the growing English language discourse around *technology*, they often remarked on how the term obscured the Continental distinction between *technique* and *technology*. This became a significant issue for Continental historians of technology in 1983 with the founding of the British journal *Technology and History*. Its first editor was Pietro Redondi, a young Italian historian of science and technology who was soon to make a name for himself with a controversial reinterpretation of the trial of Galileo (Redondi 1987).

Redondi introduced the journal with a short essay titled "History and technology: Research on the borderline" (Redondi 1983). In this essay, Redondi identified two foundations for the social history of technology. The first was the Second International Congress of History of Science and Technology in 1931, famous for the papers presented by a Soviet delegation. The second source was Lucien Febvre's 1935 paper in *Annales*, which was reprinted in the first issue of *History and Technology*, both in the original French and English translations. However, Redondi immediately raised the issue of terminology: "Should the subject be technology? Or should it be techniques?". Even though *technology* was the name of the journal, Redondi did not want to ignore the Continental distinction "between technology, designating the scientific study (the institution of which is relatively recent) of the procedures and instruments of the various branches of industry, and techniques, the aggregate of methods and procedures used in an art, a trade or an industry". However, Redondi never explained how he would preserve this distinction in a predominantly English-language journal.

Another article in *History and Technology*'s first issue also dealt with the problem of the Continental distinction between *technique* and *technology*. This article, "On the rise of technological science", was written by Jan Sebestik, a French philosopher and historian of science (Sebestik 1983). Sebestik insisted on the need for the distinction between *technique* and *technology*, a distinction that was lost in translation to English. He pointed to a growing historical literature on the concept of technology in French and German. However, he insisted that "the English speaking reader has some trouble translating (and understanding) the titles of these books. 'Technologie' is not 'technology' (which is the English world for the French technique)". To maintain the distinction in English, Sebestik proposed following Mumford and translating *la technique* and *die Technik* as *technics* while reserving *technology* for the field he termed "technological science".

Sebestik's article was essentially a plea for restoring a supposed "unitary discipline" of technology understood as the science of technics, a discipline that encompassed "systematic reflection on tools, crafts and industrial production as a specific branch of human knowledge". *History and Technology* published two more articles supporting the study of *technology* as an academic field in the journal's first two volumes (Salomon 1984; Sigaut 1985). Sebestik also helped produce an edited volume in 1984, the *De la technique à la technologie*, under the auspices of the CNRS (Sebestik 1984). However, as I argue below, there was no general discipline of *technology*. Instead, the idea of *technology* as a coherent discipline was a construct of academics seeking in part to counter the growing influence of the English-language term. Although European academics would return periodically to this idea (e.g. Agazzi (1999)), it never gained much traction, particularly among practitioners and academics working in engineering disciplines. Instead, engineers embraced the concept of engineering sciences, an idea that resonated broadly through its role in engineering education (Channell 2009).

5.3. Roots of the English-language concept of technology: an overview

The history of the term *technology* in its original meaning as a field of study played only a peripheral role in present-day meanings of the English-language concept (Schatzberg 2018, Chapter 6). Instead, the term's current meanings emerged largely through a process of mistranslation that carried the meanings of the Continental term *technique* into the English term *technology*.

In the remainder of this chapter, I tell the story of how the English term *technology* became a keyword of late modernity. The specific history of the concept of technology is embedded in a much longer story of scholarly discourse about technical activities. This discourse has a continuous history despite repeated shifts in terminology. A deep and persistent tension marks this discourse, a tension grounded in the division of labor between scholarly elites and technical workers. The historian Serafina Cuomo has called this tension "the trouble with techne". As Cuomo argues, since the time of the ancient Greeks, technical practitioners have fit uneasily into social hierarchies, especially aristocratic hierarchies based on birth (Cuomo 2007).

This tension shaped and continues to shape the discourse of technology, resulting in two divergent traditions. On the one hand, technical practitioners and their allies view technology as an expression of the human spirit, a deeply creative human activity embedded in moral values. Many scholars, in contrast, far removed from the world of the technician, have portrayed technology as a narrow form of rationality, uncreative and devoid of values, a mere instrument serving ends defined by others (Schatzberg 2018, p. 3). In between are a small set of people who bridge the worlds of scholar and technician, including scholars who immerse themselves in the worlds of technicians and literate technicians who write about their work for broader audiences (Long 2005).

However, throughout this history, the voices of the elite took precedence over the ideas of technicians. Our concept of technology is a product of the elite perspective, a perspective largely limited to the Western world and to men. There are other stories that give more agency to people outside the West, women and to other excluded technical practitioners (Bray and Hahn 2023). However, our present-day concept of technology has, for the most part, neglected these other stories. In a sense, much of the recent history of technology involves a struggle to recover these neglected stories to incorporate the experience of technical practitioners into our historical accounts. This approach is represented, for example, in the feminist history of housework (Cowan 1983) and the recent history of maintenance (Vinsel and Russell 2020) and in the rapidly growing scholarship on technology outside the West (e.g. Siddiqi (2015)).

As Redondi, Sebestik and other historians of technology noted, the English word *technology* was at odds with its etymology. It designated not a field of knowledge, discourse or study but rather the technical activities and artifacts themselves, which should be rendered *technique* in French or *Technik* in German. However, this difference in meaning between English and other European languages was relatively recent. Before World War I, English-language dictionaries invariably defined *technology* as either the science of or discourse about useful arts. The 1911 edition of the *Century Dictionary*, an impressive 12-volume work, provides a typical definition of *technology*: "that branch of knowledge which deals with the various industrial arts; the science or systematic knowledge of the industrial arts and crafts" (Whitney and Smith 1911). This definition of *technology* is very similar to contemporary definitions of *technologie* or *Technologie* in French and German dictionaries.

However, the current meanings of the English term *technology* owe little to its original definitions. Instead, its current meanings emerged largely through a process of mistranslation that carried aspects of the Continental term *technique* into the English term *technology*. This mistranslation explains, at least in part, how the term *technology* began to diverge from its Continental cognates in the first two decades of the 20th century.

However, few people noticed this divergence, mainly because *technology* was simply not a very important word in English. Before the 1960s, the term *technology* was largely confined to scholarly discourse, and even this term remained obscure until after WWII. The English term first appeared in a seventeenth-century dictionary. For the next few centuries, *technology* occupied recondite corners of scholarship, occasionally bubbling to the surface in some narrow context, only to disappear again. The term finally found a place in the consciousness of Anglo-American scholars after World War I. Even then, *technology* remained uncommon among non-specialists, except as a name for institutions of technical education (Schatzberg 2018, Chapter 10).

Not until the 1960s did *technology* become a keyword in popular discourse. Before World War II, English speakers made sense of material culture, primarily with other terms, such as *invention*, *industry*, *manufactures*, *machinery*, *science* and especially *art*. None of these terms encompassed all of what we would now categorize as technology, except perhaps for *art*. Throughout most of its history, *art* did not denote aesthetics but rather all forms of making. The concept of art remained central to discourse about material culture into the early 20th century, although it was often modified as *mechanical*, *useful* or *industrial* to distinguish it from fine art.

5.4. Discourse of the arts from the ancient world to the enlightenment

In other words, *art* was the meta-concept that encompassed what we now call *technology*. The concept of art has a long history as a keyword of Western philosophy, providing a direct link through the Latin term *ars* to the Greek concept of *techne*. In ancient Greek, *techne* referred not only to crafts but also to medicine and even rhetoric. However, the Greeks also engaged in struggles over the meaning of *techne*, struggles closely connected to the social status of technicians and their knowledge. These struggles focused on two key questions that have survived into the present. The first is the issue of boundaries. What forms of action and knowledge deserve to be classified as *techne* (or technology)? The second question concerns the moral status of *techne* and the social status of its practitioners. Are *technai* (technologies) neutral means, or do they possess inherent moral value? Are technicians virtuous or base?

These questions about moral status are key to tensions over the meaning of technology. Aristotle set the tone in ancient Greece when he made virtue in *techne* subordinate to ends defined by non-technical elites. Aristotle also articulated a hierarchy of knowledge with *techne* at the bottom, below moral and theoretical knowledge (Schatzberg 2018, Chapter 2).

Tensions over this hierarchy of knowledge continued into the Medieval and early modern eras. Most medieval authors maintained some version of Aristotle's sharp distinction between philosophical and productive knowledge, *episteme* and *techne*, sometimes framed as the division between liberal and mechanical arts.

However, this situation began to change in the 15th century in a way that partly bridged the gap between *episteme* and *techne*, between science and the arts. This shift was rooted in what Pamela Long calls the alliance between *techne* and *praxis*, that is, between technical experts and political elites. This alliance took shape as new forms of technical knowledge became increasingly relevant to political power (Long 2001, pp. 18–19). The result was a surge in authorship about the mechanical arts, including works written by both humanist scholars and artisan practitioners. Francis Bacon drew from this tradition when he began arguing, two centuries later, for a closer connection between natural philosophy and practical applications. However, as Peter Dear and Steven Shapin have shown, respect for the mechanical arts did not imply respect for the artisan. Instead, natural philosophers maintained the conceptual hierarchy of mind over hand that mirrored the social hierarchy of the philosopher over the artisan (Shapin 1989; Dear 2009).

Technology was almost completely absent from this early modern discourse. In the 18th century, European scholars increasingly focused on art, science and the relationships among them. This interest arose in part from the era's mania for comprehensive encyclopedias, which typically encompassed the full range of "arts and sciences", categories that helped frame the division of knowledge used to organize these encyclopedias. The categories of science and art were direct descendants of the Aristotelian definition of *episteme* and *techne* as disjointed forms of knowledge. However, Enlightenment encyclopedists rejected Aristotle's sharp boundary between science and art. Instead, they theorized science and art as overlapping and interacting fields of knowledge, without favoring either science or art (Yeo 2001). The most famous expression of this understanding occurs in the *Encyclopédie* of Diderot and D'Alembert, who stressed that knowledge of the arts was embodied in artisans. As Diderot confidently proclaimed: "Let us finally restore to artists (i.e. artisans) the justice that is their due" (Schatzberg 2018, pp. 58–59).

However, writers added an important distinction to the concept of art that eventually weakened its utility as a term covering technical activities. This was the concept of fine arts, introduced by Charles Batteux in his 1746 book, *Les beaux arts réduit à un meme principe*. Essentially, as Larry Shiner has argued, the concept of fine art split aesthetic creativity from "mere" craft skill so that "all the 'poetic' attributes – such as inspiration, imagination, freedom, and genius – were ascribed to the artist and all the 'mechanical' attributes – such as skill, rules, imitation, and service – went to the artisan" (Shiner 2001, p. 111).

5.5. Technology fills a semantic void

Over the next century and a half, *art* gradually became restricted to the fine arts, undermining its usefulness as a general term for material culture. As a result, by the late 19th century, no English-language term seemed suitable for describing what we would now term *technology*. This was the era of the Second Industrial Revolution, with momentous changes brought about by new industries in the field, such as electricity, synthetic chemicals and steel. In the early 20th century, *the machine* became a popular term for this new industrial complex. However, *machines* never became common in scholarly discourse. Instead, what Leo Marx called a "semantic void" arose: the absence of adequate concepts to describe the new industrial era (Marx 2010).

This void was eventually filled by new meanings of *technology* that emerged between the 1900s and the mid-1930s. During this time, a variety of scholars, mainly in the social sciences, transformed *technology* into a suitable replacement for the older term *industrial arts*. However, Marx's "semantic void" did not call forth these meanings, vacuuming them up from dusty corners in the house of language.

What motivated these changes, at least first, were intellectual dynamics within the social sciences.

The shift began when American social scientists appropriated the German discourse of *Technik*. As used by German engineers and some social scientists, *Technik* refers to the methods, tools and instruments used to create and maintain material culture. In their search for social status, elite German engineers developed a theoretical discourse about *Technik* that was explicitly cultural. Around the turn of the 20th century, the concept of *Technik* was also adopted by German social scientists, including two founders of classical sociology, Max Weber and Werner Sombart (Schatzberg 2018, Chapter 7).

However, scholars faced a dilemma when translating *Technik* into English. Sometimes they used the English *technique*, but this was a mistranslation when *Technik* was used in the sense of the industrial arts. Instead, beginning around 1900 with the work of Thorstein Veblen, American social scientists began translating the German *Technik* as "technology". By the 1930s, "technology" had become the most common translation for the industrial meaning of *Technik* (Schatzberg 2006).

While Veblen was redefining *technology* as "the state of the industrial arts", other scholars began equating *technology* with the 19th-century concept of applied science. Like Veblen's redefinition, this move represented a profound shift in meaning. No 19th-century dictionary ever defined *technology* as applied science. Technology was a science, systematic study, teaching, or discourse about the industrial arts. However, as scholars have recently argued, fundamental ambiguity exists (and remains) in the concept of applied science. It could refer to a kind of science that is suited to application. Alternatively, it could refer to the process of application itself, that is, the results of applying science. Both of these meanings were common in the 19th century. However, the use of *technology* as a synonym for *applied science* became widespread only after World War II, in part because of the prominent role played by scientists in creating the atomic bomb (Schatzberg 2018, Chapter 11).

The meanings of *technology* were also influenced by the other main definitions of *Technik*: skills and methods for achieving a specific goal. German dictionaries clearly distinguish this broadly instrumental definition of *Technik* from its definition as the industrial arts. *Klaviertechnik*, for example, means "piano technique," that is, skill in playing the piano. In contrast, *Bautechnik* means "building technology" or "structural engineering".

However, the boundary between these two meanings remained porous. It was easy to move from viewing modern *Technik* as the material expression of an era's culture (that is, industrial arts) to seeing *Technik* as technique, instrumental rationality. In the early twentieth century, Continental critics of modernity often invoked techniques to explain the dehumanizing tendencies of the industrial era. An exemplar of this literature is Jacques Ellul's 1954 book, *Technique, ou l'enjeu du siècle*, translated into English a decade later as *The Technological Society* (Ellul 1954, 1964).

When *technology* became a standard translation for the industrial meaning of *Technik*, some scholars also began translating the instrumental meaning of *Technik* as "technology". In essence, these scholars redefined *technology* as a technique. For example, Friedrich Junger's deeply anti-technological 1946 book, *Die Perfektion der Technik*, was published in 1949 in English translation as *The Failure of Technology* (Jünger 1949, 1953). Although this translation practice remained fairly uncommon until the 1960s, it further confused the concept of technology, encouraging the same combination of industrial and instrumental meanings that existed in *Technik*. However, in contrast to the German *Technik* and French *technique*, these conflicting meanings of *technology* in English were almost never clearly distinguished.

5.6. Technology triumphant

Thus, by World War II, three main meanings of *technology* had become established in scholarly discourse: *technology* as industrial arts, as applied science, and as technique. After the war, however, the tremendous cultural authority of science tended to push *technology* into the background. It was only in this era that an understanding of *technology* as the application of science became dominant. However, public awareness of *technology* in the 1950s was also encouraged by critics of modern industrial civilization. In philosophy and religious studies, scholars began to take notice of the philosophical discourse of technique on the Continent, which was translated into English as a discourse about technology (Schatzberg 2018, Chapter 12).

The rise of *technology* to the status of a keyword is clear through a simple comparison of its prevalence in the Google Books corpus (2023). *Science* has long been a central term in the discourse of modernity but not *technology*. In 1925, *technology* appeared approximately 80 times less frequently than *science*. Only in the mid-1950s did the use of *technology* begin to rapidly increase, reaching parity with *science* in 1973. From the 1980s through the 1990s, *technology* was significantly more common than *science*, often by as much as 50%, probably due to the use of the term for everything digital. However, by 2019, *science* and *technology* had stabilized at a rough parity.

Since the 1960s, the discourse of technology has become pervasive, particularly as English has become the lingua franca of global communication. However, confusion between the three principal meanings of *technology* remains common in both scholarly and popular discourse. This confusion pervades the most recent entry of *technology* in the *Oxford English Dictionary*. In a convoluted, multipart definition, the OED defines *technology* as a "branch of knowledge", "the application of such knowledge", "the product of such application", "a technological process, method, or technique", and "a branch of the mechanical arts or applied sciences" (Oxford English Dictionary 2022). In a sense, the OED captures all the principal meanings of the word *technology*. However, without awareness of the history of technology as a concept, the authors of this entry, along with present-day scholars, remain unable to disentangle its meaning.

5.7. References

Agazzi, E. (1999). From technique to technology: The role of modern science [Online]. Available at: http://scholar.lib.vt.edu/ejournals/SPT/v4n2/AGAZZI.html [Accessed 6 July 2025].

Bray, F. and Hahn, B. (2023). "The Goddess Technology is a polyglot": A critical review of Éric Schatzberg, *Technology: Critical history of a concept. History and Technology*, 38, 1–42.

Bud, R. (2014). Applied science in nineteenth-century Britain: Public discourse and the creation of meaning, 1817-1876. *History and Technology*, 30(1/2), 3–36.

Burke, M. and Richter, M. (eds) (2012). *Why Concepts Matter: Translating Social and Political Thought*. Brill, Leyde.

Channell, D.F. (2009). The emergence of the engineering sciences: An historical analysis. In *Philosophy of Technology and Engineering Sciences*, Meijers, A. (ed.). Elsevier, Amsterdam.

Cowan, R.S. (1983). *More Work for Mother: The Ironies of Household Technology from the Open Hearth to the Microwave*. Basic Books, New York.

Cuomo, S. (2007). *Technology and Culture in Greek and Roman Antiquity*. Cambridge University Press, Cambridge.

Dear, P. (2009). *Revolutionizing the Sciences: European Knowledge and its Ambitions, 1500-1700*. Princeton University Press, Princeton.

Ellul, J. (1954). *La technique ou, l'enjeu du siècle*. Armand Colin, Paris.

Ellul, J. (1964). *The Technological Society*. Knopf, New York.

Febvre, L. (1935). Réflexions sur l'histoire des techniques. *Annales d'histoire économi-que et sociale*, 7(36), 531–535.

Godin, B. (2017). *Narratives of Innovation: The Invention of Models, 1920-1980.* MIT Press, Cambridge.

Google Ngram Viewer: Science, Technology (2023). [Online]. Available at: https://books.google.com/ngrams/graph?content=science%2Ctechnology&year_start=1900&year_end=2019&corpus=en-2019&smoothing=0 [Accessed 1 June 2023].

Jünger, F.G. (1949). *The Failure of Technology: Perfection Without Purpose.* H. Regnery, Hinsdale.

Jünger, F.G. (1953). *Die Perfektion der Technik.* V. Klostermann, Frankfurt am Main.

Kranzberg, M. (1959). At the start. *Technology and Culture*, 1(1), 1–10.

Long, P.O. (2001). *Openness, Secrecy, Authorship: Technical Arts and the Culture of Knowledge from Antiquity to the Renaissance.* Johns Hopkins University Press, Baltimore.

Long, P.O. (2005). The annales and the history of technology. *Technology and Culture*, 46, 177–186.

Marx, L. (2010). Technology: The emergence of a hazardous concept. *Technology and Culture*, 51(3), 561–577.

Matschoss, C. (1909). Vorwort. *Beiträge zur Geschichte der Technik und Industrie*, 1, III–VI.

Mumford, L. (1934). *Technics and Civilization.* Harcourt Brace Jovanovich, New York.

Oxford English Dictionary (2022). Technology, n. [Online]. Available at: https://www.oed.com/dictionary/technology_n?tl=true.

Redondi, P. (1983). History and technology: Research on the borderline. *History and Technology*, 1, 1–6.

Redondi, P. (1987). *Galileo: Heretic.* Princeton University Press, Princeton.

Salomon, J.-J. (1984). What is technology? The issue of its origins and definitions. *History and Technology*, 1(2), 113–156.

Schatzberg, É. (2006). Technik comes to America: Changing meanings of technology before 1930. *Technology and Culture*, 47, 486–512.

Schatzberg, É. (2018). *Technology: Critical History of a Concept.* University of Chicago Press, Chicago.

Schauz, D. (2015). Wissenschaftsgeschichte und das Revival der Begriffsgeschichte. NTM Zeitschrift für Geschichte der Wissenschaften. *Technik und Medizin*, 23(1/2), 53–63.

Sebestik, J. (1983). The rise of the technological science. *History and Technology*, 1(1), 25–44.

Sebestik, J. (1984). Présentation. In *De la technique à la technologie*, Chrétien-Goni, J.-P. (ed.). CNRS, Paris.

Shapin, S. (1989). The invisible technician. *American Scientist*, 77, 554–563.

Shiner, L. (2001). *The Invention of Art: A Cultural History*. University of Chicago Press, Chicago.

Siddiqi, A.A. (2015). Technology in the South Asian imaginary. *History and Technology*, 31(4), 341–349.

Sigaut, F. (1985). More (and enough) on technology! *History and Technology*, 2, 115–132.

Singer, C., Holmyard, E.J., Hall, A.R. (eds) (1954). *A History of Technology*. Oxford University Press, New York.

Vinsel, L. and Russell, A.L. (2020). *The Innovation Delusion*. Currency, New York.

White, L. (1940). Technology and invention in the Middle Ages. *Speculum*, 15, 141–159.

Whitney, W.D. and Smith, B.E. (1911). *The Century Dictionary and Cyclopedia*. The Century Co., New York.

Yeo, R. (2001). *Encyclopaedic Visions: Scientific Dictionaries and Enlightenment Culture*. Cambridge University Press, Cambridge.

6

Industry in the History of Techniques

Guillaume CARNINO
COSTECH, Université de technologie, Compiègne, France

6.1. Introduction

Industry is not just a modern European phenomenon, as evidenced by the buried army of the first emperor Qin Shi Huangdi, whose tomb, covering an area of over 56 km^2, required a workforce of over 700,000 and contains several thousand statues of horses and soldiers produced in series, although individually differentiated (Ciarla and Luca 2005). As early as the 4th millennium BC, the city of Uruk stood out for the "remarkable number of industries that developed in these temple workshops, as documented in the cuneiform accounts. Among them, we find the first large-scale dairy and wool production; also the manufacture of leavened bread, beer and wine, including facilities for standardized packaging" (Graeber and Wengrow 2021, p. 350). The authors also mention the existence of genuine quality control carried out in these urban temples, where a cylinder seal was affixed to manufactured products to mark their provenance and authenticity. Even if archaeologists working on Archaic Greece do not use the term "industrial", the manufacture of votive figurines reached considerable proportions, with mass-produced models found throughout the Corinthian area in the 7th and 6th centuries BC (Bilbao 2020). In Latin Antiquity, iron bars were standardized according to well-defined benchmarks throughout the Roman Empire, sometimes produced by proletarianized slaves in concentrated workshops (Pagès and Verna 2022). The amphora itself, a genuine Roman "tin can", was a widely standardized container for the distribution of edible products in the Empire (Pagès and Verna 2022). A study of the stamps affixed to Roman bricks

History of Techniques,
coordinated by Guillaume CARNINO and Xavier GUCHET.

shows that this industry (of which mass storage was an integral part, forming the basis of a genuine regional economy) was owned by the most important aristocratic families and members of the imperial court (Thébert 2000). Even in the Middle Ages, some metallurgical production was carried out on an industrial scale – a situation that was far from exceptional, as historian Catherine Verna shows for the case of Arles-sur-Tech in the Pyrenees (Verna 2017). During its European golden age, the sailing ship was a place of experimentation and industrial pretraining: "The disciplinary methods designed to transform brigades of human beings into well-oiled machines were implemented on merchant ships before being transposed to factory workshops" (Graeber and Wengrow 2021, p. 517). This is evidenced by the significant proletarization that took place (Rediker 2010), the existence of a gigantic technical supply system for naval construction and outfitting (wood, tar, canvas, powder, metal (Dessert 1996)), and the deployment of commodity capitalism of the first globalization, whose socioanthropological impact extends far beyond the territories of coastal cities (Hilaire-Pérez 1996), and even the etymology of our contemporary French concept of "panne" referred to the orientation of a ship's yards to halt its course – a controlled halt, essential for controlling the vessel, which came to mean a cessation of operation altogether (Carnino and Marquet 2020) and *risk* (from the 14th-century Italian *risco*, the possibility that a merchant ship or sailor might not return to port (Bernier 2007, p. 128)).

It is even possible that the first industrial era emerged at the same time as the empires of Antiquity. Thousands of ceramic molds (used to bake the *bedja* bread consumed by the workers) found at Gizeh argue in favor of this thesis (Graeber and Wengrow 2021): the first "mega-machines" described by historian Lewis Mumford would have been made up of human "cogs" (Mumford 1974). Of course, too many factors were lacking at the time to allow the emergence of a genuine industrialization process (mass transport allowing worldwide circulation of raw materials, separation of scholarly activity from moral thought, individualization of inventive property, complete monetary abstraction, etc.). Does this mean that there is no difference between ancient and 21st-century production? No, of course not, but to grasp the specificity of the concept of industry, we need to take the diversity of its historical meanings and contemporary incarnations seriously, which means insisting on the multiplicity of intersecting and parallel temporalities that are impossible to fit into a one-dimensional fresco.

To encompass the main areas of contemporary historiography, we first attempt to define industry and then set out to uncover the diversity and multiplicity of the forms of its growth over the last two centuries before ultimately focusing on the question of expertise and its machine integration.

6.2. Defining industry

While the notion of industry is part of everyday language, the multiplicity of industrial contexts calls for a more refined conceptual definition. The Latin *industrius*, referring to activity, zeal and assiduity, is formed from the two radicals *indu-* (meaning "into" or "within", in the manner of the prefix endo-) and *struere* ("to gather", "to build", "to weave", or more broadly "to make", from which today's verb *construct*, for example, is derived). With its many meanings, *industria*, as the ability to "make things", is simultaneously fabrication, skill and cunning. While industry may have been production or even machinery, it has also symbolized cunning and the misuse of legitimate means, such as the clever genius in *Méditations métaphysiques* who employed "all his industry"[1] to deceive René Descartes. Here, we can find the etymology of many terms relating to technology that combine machines and machination, production and cunning, weaving of materials and interweaving of humans (to that mythological point where the demiurge is already a potential trickster). Until the 18th century, industry referred to the use of unfair means to achieve economic objectives (Fontaine 1992) while at the same time being a fully-fledged synonym for the notion of economy (Fontaine 1993). Moreover, the ancient meaning of industry does not establish any disjunction with the notion of craftsmanship.

It was in the midst of industrialization that industrialists and craftsmen came to be pitted against one another: during the 1930s, the specific category of craftsmanship was established in Europe as the other category of industry, an opposition endowed with a strong symbolic value. Oppositional pairs of terms are thus sedimented, redoubling the allegedly structural antagonism between the artisanal and industrial worlds: manual labor versus mechanization, artisanal specialization versus industrial mass production, unity of tasks versus division of labor, independence versus proletarianization, animal or human energy versus mechanical energy, traditional methods versus innovative processes, local production versus the global market, regional specificities versus standardized quality, or even technique versus technology.

Each of these canonical oppositions can be used to qualify dated and identified productive phenomena: this is the case with leather, which in the 18th century was obtained using vegetable tannins at the end of a process lasting 1–2 years and is in no way comparable to the chrome tanning carried out today in a single bath lasting

---

1 "Sed est deceptor nescio quis, summe potens, summe callidus, qui de industria me semper fallit" ("But there is a very powerful and cunning deceiver, who uses all his industry to always deceive me") (Descartes [1641] 1992, pp. 72–73).

approximately 24 hours (Baptiste et al. 1998; Millet 2012). Similarly, the manufacture of the surfboard in Hawaii before the arrival of European settlers is a highly symbolic act of individuation and identity, linked as it is to the social prestige of its owner (Lemarié 2018), whereas in 2018, the surf equipment market – estimated at 3.1 billion dollars – is based on the polymer industry (Grand View Research 2019). While these examples seem to support the existence of a clear difference between craft and industry, whether in terms of production methods or market scope, the situation is in fact infinitely more complex: on the one hand, the two terms of the previous oppositions often overlap and combine; on the other hand, these oppositions themselves do not overlap (manual labor can be used to produce for a global market, traditional methods can feed the division of labor, animal power can fuel mechanization, etc.), so much so that it is necessary to challenge these caricatured antagonisms.

Regardless of the criterion (or pair of opposites) chosen, it is always possible to find counterexamples. If industry is to be defined by the use of machines, we need to consider the fact that the manual (artisanal?) production of hats in Chile has reached industrial proportions (Flacher and Pelletan 2007), or conversely, many machines are used in craft model-making (airbrushes, drills, 3D printers, etc.). If we defend the division of tasks as the basis of any industrial process, then we need to explain how operations could have been so widely segmented in the luxury crafts of 18th-century London to the point where a genuine technical grammar was developed for keeping accounting records (Hilaire-Pérez 2013). If we take market size as the primary criterion, we must avoid describing the development of fish farming as industrial, despite its being called "industrial" under the Second Empire, whose model establishment in Huningue, at the origin of its expansion, was entirely financed by the government and supplied all the other provincial sites free of charge (Carnino 2014a). Innovation, which was thought to drive industrial development, is in reality by no means the prerogative of capitalist entrepreneurship, given that artisanal communities have always practiced collectively modeled and controlled invention (Hilaire-Pérez 2000). Attaching industry to the transformation of raw materials (a semantic link found as far back as 1771 by Abbé Baudeau ([1771] 1910), reinventing the categorization of the three economic sectors – primary, secondary and tertiary – stemming from American economics and still taught in high schools (Flacher and Pelletan 2007) does not resolve the issue either. This is because, on the one hand, craftsmanship is justly deserving of this qualification, and on the other hand, there are sectors today that we would readily describe as industrial, yet do not transform any raw materials into manufactured products: we are thinking here of logistics platforms, which have undergone massive industrialization over the past few decades (to the point where the sector has now overtaken the construction industry – though historically the leader – in terms of occupational illnesses and accidents; Benvegnù and Gaborieau (2017)), or of call centers, where managers have learned to manage temporary workers right down to their smiles (Amiech 2003), industrialized agriculture or mining (which should be the heart of the primary sector) or even of datacenters, which do not produce anything physical beyond the size of an electron yet mobilize infrastructures typical of heavy industry (titanic fuel tanks, giant air conditioners, SEVESO-classified facilities, etc.; Carnino and Marquet (2018)).

We can, of course, decide to use an arbitrary criterion to define industry, but then, many borderline situations escape common sense. For example, datacenters, which were originally classified as tertiary since they were part of automation and the IT sector, have raised questions in several French local authorities, as the ground resistance and ceiling height of the buildings have sometimes tipped them over into the industrial sector, which can lead to administrative contradictions, not least in terms of taxation (Carnino and Marquet 2022). In a similar vein, one of the demands of local authority tax departments over the past decade has been to review the tax status of logistics platforms to no longer depend on case law, a political effort that led to the redefinition of what makes an industry in the French tax code in 2019:

> Buildings and land used for manufacturing or processing movable tangible goods requiring substantial technical resources are considered industrial.
>
> Buildings and land used for activities other than those mentioned in the first paragraph of A, which require substantial technical resources and where the technical installations, materials and tools used play a predominant role, also have an industrial nature.
>
> However, in the two cases mentioned before, when the value of the technical installations, equipment and tools present in the buildings or on the land and intended for the activity does not exceed an amount of 500,000 euros, these buildings and land do not have an industrial nature.[2]

Industry is initially defined in a vague way (manufacturing, processing or other activity) before the real criterion of qualification is specified: the size of the technical resources (which, in this case, must necessarily exceed 500,000 euros). This gives us a clear idea of what lies at the heart of the industrial phenomenon:

2 General Tax Code, Article 1500. We would like to thank Clément Marquet for this valuable reference.

industry arises when a technical phenomenon grows in quantity to the point of changing its nature. Industry is nothing other than a *large quantity of technical means* grouped together in time (home production, subcontracting) or space (manufacture, factory). For each of the abovementioned criteria and oppositions (market development, workforce concentration, task segmentation, innovation, standardization, etc.), the question relates to the growth threshold at which the very nature of technical activity radically changes: "that which changes scale changes essence", as anthropologist Michael Singleton reminds us (Singleton 2007). While raising one pig may raise hygiene issues, fattening a thousand pigs creates environmental problems (Espagnol and Leterme 2010). Setting up a text search in an online database can be achieved in a few minutes for a competent developer; making (almost) the entire Web searchable is a colossal undertaking, requiring more than a million servers (and the associated energy and environmental costs) and is now called Google, Bing and Yahoo. Every technical act – whether peeling a carrot, pressing the "power" button on a computer, knitting a sweater or answering the phone – raises new questions and requires new infrastructures when carried out on an industrial scale.

Artisanal communities have always innovated, but when this technical aspect is transferred to enormous research and development centers, we fully enter into an industrial process: the American electrical grid emerges from Thomas Edison's Menlo Park, which brings together an immense scientific library, machine tools, means of transport, a chemical laboratory, steam engines, a library of existing patents, etc. (Hughes 1989). As early as 1876, Edison himself spoke of an "invention factory" and predicted that his team would come up with a minor invention every 10 days and a major one every 6 months: the industrial growth of the invention process changed its pace and, *in fine*, its nature.

At the beginning of the French Third Republic, Louis Pasteur suggested that silkworm farmers use microscopes to detect the presence of diseases (such as pebrine or pepper disease) upstream of silkworm production. At the Milan Congress in 1876, the scientist was satisfied with seeing a factory (whose front was adorned with his surname) operating according to his method, employing 70 female "micrographers" to examine 40,000 butterfly cells a day (Carnino 2014b). The work of the scientist in the laboratory examining samples of mulberry bombyx is used as a matrix but is not of the same nature as the activity of the Milanese women workers: sericulture has changed scale, and it has been industrialized.

Historians Giorgio Riello and Prasannan Parthasarathi have studied the development of the cotton industry and see two sets of theses explaining Europe's success at the expense of India and China (Riello and Parthasarathi 2009): the "quantitativists" stress economic and institutional factors, whereas the "qualitativists" emphasize technological leaps[3]. Thinking of industry as the quantitative growth of a technical phenomenon to the point where we witness a qualitative break makes it clear that the two points of view are by no means irreconcilable and, once again, that causality in history is invariably multifactorial.

Each of the aforementioned oppositions between artisanal characteristics and industrial attributes is, in fact, a progressive gradient – if not a linear one – which sometimes crosses thresholds and generates an essential transformation. We can also question the links between these different antagonistic pairs as the technical scale increases: quantification can sometimes lead to the deskilling of tasks – the Taylorist specialized worker embodying this opposition to the skilled worker and master of specific expertise.

Defining industry as what happens to a technical phenomenon when its quantitative growth leads to its qualitative transformation was essential to frame the discussion. However, this does not solve the historical problem of the advent of our contemporary societies, which are sometimes mistakenly seen as the result of an alleged industrial revolution that changed everything overnight. As we shall see, this view is erroneous, and it is more appropriate to speak of a process of industrialization.

6.3. "Industrial revolution" and industrialization

Today, the mythology of the Industrial Revolution is taught from an early age. The "*Mes p'tits Docs*" children's collection sold by French publisher Milan is a good example of this, since its book, entitled *La Révolution industrielle* (The Industrial Revolution), opens as follows: "It all begins in 1769 in England: engineer James Watt develops a machine that will change the world" (Ledu and Germain 2016). To condense so many approximations and historical errors into a single sentence is quite a feat: the history of the steam engine predates 1769 by a long way, since Thomas Savery filed a patent for a drainage pump as early as 1698, before teaming with Thomas Newcomen in 1712 to produce a "fire engine" also intended for sumps; although England has a staggered industrial timeline (notably due to its insularity, early deforestation and the importance of its carbon resources), industrialization is by no means a matter of British history alone; James Watt was not an engineer but an instrument maker (who did not invent but only repaired and

3 Maxine Berg and Pat Hudson had already pointed out the distinction between quantitative (gradualist) and qualitative approaches, encouraging re-readings of industrialization (Berg and Hudson 1992).

improved a miniature steam engine for the educational needs of a university sponsor); last but not least, the steam engine did not "change the world", it was painstakingly integrated into it, and was initially more a consequence than a cause of capitalist and industrial dynamics. Nevertheless, this mythology is so strong that some authors have gone so far as to number the various industrial revolutions in history: most often, the first refers to the coal and steam revolution (England, 18th century), the second to the chemical and electrical revolution (Europe, second half of the 19th century), the third to the telematics and nuclear revolutions (1950s and after (Caron 1997)) and the fourth[4] to the NBIC convergence (nano-, bio-, info- and cogno-technologies) driven by digital technology (although some refer to this contemporary phenomenon as the third industrial revolution). Thinking about industry historically requires us to distance ourselves from the mythology of invention that underpins these clichés about industrialization (and which has contributed to the construction of today's industrialist dead ends).

As historian Jean-Baptiste Fressoz noted:

> The idea of an "industrial revolution" was born in France among the learned elite who had emerged from the revolutionary wars. The expression refers to the fight it intended to wage against England: France was to oppose an empire of sailors and merchants with an empire of industrialists and scientists. [...] It refers to a series of technological successes linked to the blockade and autarky (extraction of beet sugar, synthesis of sodium carbonate, etc.), and is based on an exacerbated confidence in technological fiat, foreshadowing a sudden transformation of the productive system that would enable France to catch up with England almost instantaneously. (Fressoz 2015, pp. 370–372)

This "industrial revolution" entered the economic and social vocabulary of the 1830s to describe the reconfiguration of society and commerce, following the introduction of steam engines in the textile industry (Vincent 2021). In the 1880s, lectures by the economic historian Arnold Toynbee popularized the expression as a critique of the capitalist market economy (Kadish 1986) before his students turned it into a purely analytical and descriptive object of historical development (Jarrige 2022). The second half of the 19th century, with its emphasis on science and progress, soon saw the invention of mechanics as the key to the modern world, and a heroic vision of the Industrial Revolution gradually took hold until the end of the

4 *L'Usine nouvelle* headlines: "South Korea aims for its fourth industrial revolution". Available at: www.usinenouvelle.com/editorial/la-coree-du-sud-veut-faire-sa-quatrieme-revolution-industrielle.N612958.

20th century, when economic historians challenged this simplistic perception (Berg 1985; Verley 1997).

They discovered that the spread of the steam engine throughout the world was slow and highly variable, that it was generally preceded by various processes of industrialization (segmentation of tasks, proletarianization, etc.), and that it often coexisted with other engines (primarily hydraulic but also animal or human; Inskter and O'Brien (2004)). As historians François Jarrige and Liliane Hilaire-Pérez note, "while British industry used 30,000 steam engines around 1830, there were no more than 1,000 in France at the same date, and only 6,000 in 1852. [...] The final choice of the steam engine had less to do with its intrinsic superiority than with its greater flexibility" (Hilaire-Pérez and Jarrige 2021, p. 29). Returning to the canonical narrative, historian Andreas Malm shows that the steam engine was globally counterproductive from an economic and technical point of view (Malm 2016): hydraulic mills were better controlled and more reliable than boilers, the latter being sources of explosions and unpleasant odors. However, coal-fired machines had an important cultural advantage at the time: they enabled entrepreneurs to free themselves from geographical constraints (waterways, structured and militant employment pools) and temporal constraints (dry season, working hours accepted by workers) in favor of production that was riskier in some respects but less collective and more controlled from a hierarchical perspective. The steam engine made it possible to ignore territorial, seasonal and cultural constraints, giving free rein to entrepreneurial competition and a liberal spirit. As Karl Marx had already described in *Capital* (Marx [1867] 1993), it is therefore logical that proletarization and the gradual takeover of production by merchants, to the detriment of tradesmen (a trend whose beginnings can be traced back to the Middle Ages), should have preceded mechanization and the introduction of the inanimate motor (Halleux 2009).

The steam engine is not, therefore, the miracle instrument that created the contemporary world but rather the technical crystallization of capitalist social relations. To put an end to the old renewable technical forms based on living water (Benoît 2020) and animal energy (Jarrige 2023), we had to resort to a new regime derived from geology's long history – coal, the combustion of which provided rapid access to large energy reserves. In other words, major industrial transformations took place without major technical changes, and the steam engine merely cemented the process and provided a new impetus. Once and for all, however, we must be careful not to overplay its historical role, as the historiography of recent decades has amply demonstrated the diversity of industrial deployment processes.

Political scientists Charles Sabel and Jonathan Zeitlin highlighted the many forms of industrialization (Sabel and Zeitlin 1985). Questioning historical alternatives to mass production, they challenge technological teleology and show the multiple paths taken by industry, which are often intertwined: small units, home-based work and vast networks of subcontractors coexist with large groups, concentrated workshops and mechanized production lines, from north to south and from yesterday to today. Numerous historical studies have highlighted the complementarity – rather than the opposition or inevitable succession – between these typical ideals: *Kaufsystem* (when producers and traders contract standardized goods without permanent dependence, under the quality control of municipal authorities, e.g., Cerma and Ogilvie (1994)), *Verlagssystem* (based on decentralized home working; Pfister (1992)), manufactory (a more centralized establishment where production is done by hand, as in the *alla milanese* silk twisting industry, made with wrist-powered mills; Caspard (1979)) and factory (a mechanized workshop concentrating labor and productive capital; Hatzfeld (2022)). As the historian Maxine Berg (1980) noted, "The traditional crafts and manual labor might continue, but they now did so within the circuit of industrial capital" (p. 2). Throughout the 20th century, many globalized goods, such as popular jewelry (Carnevali 2003), English cheeses (Blundel and Tregear 2006) and Japanese watches (Donzé 2011), have integrated different modes of production throughout the chain at various levels.

François Jarrige reminds us that these long-hidden analyses are by no means new, as demonstrated by the work of historian Henri Hauser at the dawn of the 20th century, who denounced two pitfalls of economic history (Hauser 1927):

> Over-valuing exceptions – such as believing that English drapery was already a concentrated industry, because there were a few factories in the monasteries secularized under Henry VIII – or conversely forgetting the persistence of ancient forms in the present. 'In the France of 1917, alongside factories employing thousands of workers, do we not note the survival of small trades, which are very little different from those we know from ancient texts?' […] Significantly, [Hauser] refers to the United States, a symbol of novelty with its large automated factories, but which have not made small shops disappear. […] For [him], 'things appear more complex and fluid to the historian's eye', and the historiography of the last thirty years, focused on the many forms and paths of industrialization, adds nothing to this first observation. (Jarrige 2022, p. 277)

This complexity of the industrialization process has been reinforced by a review of the role played by craftsmen (Berg 2014), women (Berg 1993) and children (Humphries 2003). The recent summary by historian Barbara Hahn shows the extent to which multifactoriality prevails, to a very large extent, in the dynamics of industrial deployment, which sometimes mobilizes artisanal virtuosity within mechanized production units (Navel 1979; Hahn 2020).

Logically enough, all this work has brought back the question of choice in the industrialization process, and this reintegration of politics into the old neo-liberal narrative has completed its disruption. Nevertheless, the emphasis has sometimes been so much on these socioeconomic factors, describing industrialization as a continuous process of recomposition between actors, firms, markets and states that the technical aspect has been overlooked (Sabel and Zeitlin 1997; Herrigel and Zeitlin 2010). While historian Philip Scranton insists on the plurality of production regimes (in 1907, specialized production and small series accounted for at least one third of the economic value produced in the United States, and during the first quarter of the 20th century, mass production did not exceed 40%), he does not overplay the social determinants of this industrial history. The obsession of the engineers and entrepreneurs he studies was not politics but the technological knowledge and expertise that underpinned their shared identity: minute technical variations could make all the difference between commercial success and disastrous failure (Scranton 1997). The point here is not to reintroduce the linear industrialist narrative thrown out by Sabel and Zeitlin but rather to understand that industrial flexibility has its limits and that the whole art of manufacturing consists of playing with technological constraints.

Notably, while textiles may not have triggered a revolution in production as was once thought (Rioux 1971), there is nonetheless a constitutive hybridity to clothing, the second human skin central to the production of the self and the world, which always involves blending disparate elements into a coherent whole with a defined form. The etymology of a technique such as weaving (from ancient Greece, even Indo-European) prefigures industry as a multifaceted phenomenon, taking a wide variety of paths. As historian Lisa Caliste noted, "drapery, in its broadest sense, in other words, the production of woolen fabrics for clothing, was unquestionably the leading craft and industrial activity in the Middle Ages" (Caliste 2021, p. 45). The fact that clothing production is – according to our contemporary categories – both artisanal and industrial highlights an analytical difficulty: our concepts lock history into anachronistic models. Following the footsteps of economist Joseph Schumpeter, innovation is all too often thought of in terms of disruption, even though this is a rare occurrence, and that industry generally emerges owing to incremental innovations that gradually recompose – sometimes completely – the production process. Medievalists Philippe Bernardi and Catherine Verna noted how difficult it is for the model of disruptive innovation to describe the complexity and multiplicity of developments in metallurgy and drapery over several centuries, proposing to describe this repertoire of technical transformations as microinnovations or common innovations (Benoît 1998; Bernardi and Verna 2001). In a similar vein, historian Joël Mokyr emphasized the abundance of invention since the 18th century (Mokyr 1990), when many seemingly anecdotal developments, such as John Wilkinson's cylinder-boring machine, which is crucial to Watt's machine, proved decisive. The recently compiled works of the contemporary-period scholar Serge Benoît show that the industrialization process often took place in the first instance, using improved old techniques or local adaptations (addition of piston blower machines, circular blast furnaces, etc. (Benoît 2020)). Wood-fired metallurgy, which reached its apogee in France in 1856, was thus long maintained in Alpine territories (Judet 2019) – therefore invalidating the evolutionary chronology that depicts the history of techniques as a succession of distinct phases. The work of historians and anthropologists, especially Dewerpe on proto-industrialization in northern Italy in the 19th century, places just as much emphasis on the dense, organized constellation of interrelations between town and country, between agriculture and industry, and between the old urban artisanal space and the new rural productive space (Dewerpe 1985): we are a long way from the pure concentration of factories, and we understand how industrialization is a multifaceted process affecting many different economic and social spheres. Similarly, economic historian Patrick O'Brien has shown that major textile inventions do not follow one another according to the challenge–response model but rather lead to improvements and new products within a highly skilled mechanical environment that underpins the industrialization process (O'Brien 2012).

If there is a disruption, it is, in fact, much more the trace of the archive's irruption into accessible documentation than the mark of a disruption caused by innovation, which does not proceed by leaps and bounds (Verna 2017). On the site of the Zürich town hall, built between 1898 and 1901, different generations of techniques coexisted (Rauhut 2018): workers had a ramp and a lifting machine at their disposal, and if they used both simultaneously when there was a large amount of material to transport, the weighting of operating costs, the height to be reached, the available workers and the parts to be assembled often made it possible to mobilize one rather than the other. A retrospective illusion leads us to see a disruption in the chronology where, in reality, technical generations usually cohabit on a long-term basis. For example, historian Géraldine Barron shows how, in the world of shipping, sail and steam have for a long time been more closely associated than opposed (Barron 2022). Even in our contemporary consumer world, where obsolescence is programmed as quickly as possible, we only need to look at the contents of our kitchens to notice the many technical timeframes that coexist there: oyster knives and electric mixers are no more opposed than cloth napkins and rolls of paper towels. Certain uses are gradually being redesigned, but it is rare to hand in one's favorite corkscrew the moment another, supposedly more sophisticated, is offered.

We can see that industrialization is a gradual transformation, which must initially fit into numerous dynamics before recomposing them in return – as in the case of oil, initially mainly used as a lubricant (more than as a fuel), a relationship that gradually reversed with the massification of new combustion engines at the dawn of the 20th century. Anthropologist Anna Tsing (2012) showed how sugarcane export practices in the colonial context of the 18th century transformed the economic, social and even environmental landscape of plantations and then that of the whole world. The adaptation required to make the first plans work on Madeira led to a transformation of colonial agricultural practices.

The automotive industry is a perfect illustration of this dynamic. According to economic historian Patrick Fridenson (1972), "until 1914, the factory was no more than a group of workshops characterized by a high degree of independence for the supervisor and initiative for the worker" (p. 575). The drive to maximize production led to a gradual but consistent recomposition of the work performed to such an extent that it ended up changing its very nature. As labor historian Nicolas Hatzfeld shows, the industrial growth of the Peugeot-Sochaux site led to a decrease in the time allotted for each operation, which in turn induced geographical fragmentation and greater control over the entire chain. In 1956, this growth was matched by the arrival of conveyor belts (the famous "assembly lines"):

> The timed arrival of mechanical components on final assembly lines imposes an unprecedented rigidity of synchronization on mechanical and bodywork factories, which is unique in the automotive industry: each engine corresponds to a precise body, which must be in its place on an assembly line when it arrives. As a result, mechanical and assembly, sheet metal and paint plants lose a lot of their technical autonomy. They tend to become simple workshops, integrated into the orchestrated flow of production, which is becoming the new definition of the factory, rather than the former specificity of the different manufacturing processes. Final assembly becomes the heart of this new whole, designed by the general conveyor, whose circuit becomes more complex as new workshops are built. (Hatzfeld 2022, p. 300)

Adapting conveyors to the factory as it existed implied transforming it to such an extent that it ultimately had to be dismantled: the "factory" thus became an apparently heterogeneous spatial whole (integrating the temporalities of suppliers, as well as workers), but, in reality, it precisely coordinated with the aim of optimizing the rate of production (Perrot 1983). Although the globalized factory can now produce at an accelerated pace, the world has, in turn, had to be largely recomposed to achieve this industrial feat.

In his famous study of the "takeover" by mechanization, Siegfried Giedion showed that the rapid maturity of automotive production processes was due to the techniques that were previously developed in slaughterhouses:

> If the automotive industry, for example, was able to develop its own assembly-line system so quickly, it is thanks to the meat industry's long habit of working on a moving object. (Giedion 1980, p. 210)

Similarly, the history of meat production shows that the industrialization of supply largely preceded demand, to such an extent that enormous quantities of prime animal materials were, for a long time, discarded: the heads, exposed ribs, collars and spines of hundreds of thousands of pigs on their way to Chicago's slaughterhouses were thrown away (Giedion 1980, pp. 198–199). In the 19th century, potential agricultural and livestock production areas in the United States were incomparable to those in Europe, while cities were still of comparable size. Cereal crops were highly productive, and rather than being reduced, their use was diversified, so that surplus production could be absorbed artificially, resulting in the overproduction of other products, such as pork, the surplus of which was thrown away. This economic gamble is based on the fact that what is hard to sell today will sell tomorrow and that demand will gradually follow supply. This logic has driven the intensification of industrialization in a number of sectors over the course of the century, and once demand has taken off, there has been no turning back.

The rewriting of the history of consumption in recent decades (Chessel 2012) has shown that changes in production are not the only factor in the process of industrialization, but that there is a simultaneous increase in consumerist practices, as modest families sometimes attempt to access certain goods that were previously reserved for aristocrats or the bourgeoisie (Gerritsen and Riello 2016; Bruland et al. 2020). Parallel to this work, historian Jan de Vries has attempted to go beyond the history of consumption to show that an increasing number of households intensified and increased their working hours to multiply market-oriented activities to the detriment of those aimed at self-consumption – working more to consume differently (de Vries 2008). This phenomenon, which has been intensely debated in historiography (Herment 2017; Moreno 2017), has been described by de Vries as an *industrious revolution*, borrowing the expression coined by historian Akira Hayami to describe the method by which Japanese farmers increased their production to survive in the market economy from the 18th century onward (Hayami 1986). The intensification of work and the proven growth in rural productivity were seen as additional drivers of long-term industrialization.

This is not to say, however, that industrialization was a long, quiet river, with countless hard-working boats sailing in harmony. In the 19th century, factory production was constructed through the thoughtful implementation of disciplinary measures (Foucault 1975). Historians Jelle Versieren and Brecht De Smet emphasize the "strategies of fear" implemented by employers to maintain a patriarchal hold over workers, whether in home-based subcontracting or in concentrated manual labor (Versieren and De Smet 2019). This tight control of productive spaces and activities largely coincided with liberal strategies: the employment of intermediary bodies very much feared by the local workforce (such as skilled workers, foremen and engineers poached from abroad), the hiring of entire households under the leadership of the *pater familias* responsible for enforcing productivity requirements in their own way, and the use of severe fines or corporal punishment in the event of non-compliance – all this instilled a culture of subordination, enabling labor discipline to be maintained.

In return, employees opposed wage deductions or harmful working conditions by leaving their posts without notice, individually or collectively, producing defective goods or sabotaging production tools. Others sometimes directly attacked machines, which materialized the balance of power within the factory and embodied the employer's stranglehold on the pace and process of production (Jarrige 2014).

Adam Smith, whose writings largely precede industrial institutionalization, was not the enthusiast of liberalism that many imagine him to be today. While he condemned the dependencies that existed between journeymen and masters or merchants governing hiring (Smith [1776] 1976), he also insisted that the individual autonomy that could be fostered in a world of small-scale producers and manufacturers also carried a risk of social disintegration, owing to avaricious capitalists directing workers who were treated like mindless animals – hence, for him, the need for a constituted moral economy, which was indispensable in a world where urban councils and governments prioritized the interests of elites (Pack 1991).

Finally, while industrialization has grown globally over the past two centuries, this does not mean that local deindustrialization processes have not occurred, as production activities are often transferred to other areas or regions (Fontaine and Vigna 2019; Raggi 2019) – to the point of increasing wage pressure on certain sectors that still exist in our areas, for example. In her analysis of the globalized textile industry, sociologist Christa Wichterich even uses the metaphor of a manufacturing ship dropping anchor in countries where the workforce is low paid and poorly organized before setting sail again for greener pastures when workers become too demanding or if more restrictive labor laws are instituted (Wichterich 1999).

For now, let us remember that industrialization is a multifactorial process that does not proceed smoothly and that its timeframes vary according to sector and geographical area. To close out this historiographical journey, we present a number of works that highlight the importance of the formalization of know-how in industrial history.

6.4. Know-how, grammatization and proletarization

The transition from a world based on artisanal production to today's predominantly industrially produced world involves the integration of artisanal know-how into industrial production chains, whether these are concentrated in the factory or dispersed in vast subcontracting networks (the advent of this fundamental reorganization of knowledge and production is detailed in Halleux (2009)). Industrialization (and its corollary: proletarization as artisanal dispossession (Polanyi 1983; Thompson 2012)) was only made possible on a massive scale by the science that came about during the European 19th century. This science, far from being speculative, is deeply rooted in reality: it is an act of *doing*. Its vocation is to produce facts and processes from laboratory instruments, whose concerns mainly overlap with those of industry, and then to transfer these processes to production units. This science that was invented and deployed in a way that was both socially and lexicographically unprecedented (what historian Harry Paul has expressed as the emergence of the "French scientific empire" (Paul 1985; Cunningham and Williams 1993), closely linked to the unprecedented industrial boom of the Second Empire (Beaud 1995; Daviet 1997)), was rooted both in an emerging scholarly and academic environment and in the industrial concerns that generated unprecedented experimental dynamism (Carnino 2015). Although natural philosophy was once often practiced by a few monks or aristocrats interested in astronomy, botany or alchemy, science, as it unfolded during the 19th century, became the mass production of facts from machines and calibrated devices (Pestre 2003; Conner 2011).

In France, the most emblematic figure of this new alliance between production, science and the state is undoubtedly Louis Pasteur, whose entire career was spent in the service of industry (vinegar-making, brewing, viticulture, sericulture, livestock farming, etc.), albeit in the name of *pure science* (Carnino 2014b): science was therefore in no way "perverted" by its industrial concerns, as it was the latter that provided it with its material, its questioning, and sometimes even its solutions. The emergence of "science" corresponds homothetically to industrial development. This "scienceization" is a further important step in the process of dispossessing know-how: it is no longer just a question of entering workshops to understand what is being done there to organize or tax them but of completely replacing craft skills (too local and dependent on contexts and individuals) with potentially relocatable industrial devices managed by scientists and entrepreneurs. In other words, the advent of science in production heavily affects the balance of power between labor and capital.

At the turn of the century, Taylorism drew on the prestige of science to naturalize an approach that empowered managers (Braverman 1998), marking yet another milestone in the history of industrial proletarianization. The American engineer Frederick Winslow Taylor intended to scientifically organize work by dividing it horizontally (segmentation of tasks) and vertically (strict separation of the roles of the engineer who directs and the worker who executes). According to Taylor (1919), the manager must assume the burden of "gathering together all of the traditional knowledge which in the past has been possessed by the workmen and then of classifying, tabulating, and reducing this knowledge to rules, laws, and formulae" (p. 36). Actions are timed and dissected by the engineer, who adapts tools to optimize production by eliminating unnecessary downtime. Taylorism definitively aims to remove expertise from workers who have become "specialized". As economist Benjamin Coriat points out:

> The effectiveness of the job as a means of resisting the intensification of work lies in this simple fact: knowledge and mastery of industrial operating procedures are, in principle, the exclusive property and monopoly of the working class. This monopoly is certainly "fractioned", "serialized" between the different professions, but for all that – and this is the essential point – the bosses are excluded from it. (Coriat 1982, p. 45–46)

This scientific appropriation of expertise marked a decisive turning point, which the workers were clearly aware of: in 1913, the *Union corporative des ouvriers mécaniciens* (Corporate union of mechanical workers) criticized the situation of "the worker reduced to the state of a brute, forbidden to think, to reflect, to the state of a soulless machine, producing intensively with excess, until premature wear and tear, making him a non-value, gets him dismissed from the workshops"[5]. Even workers' resistance serves as inspiration for new forms of grammatization (Auroux 1994;

5 Leaflet from the Renault Strike Committee, during the February 1913 strike against timekeeping (reproduced in Hinrichs and Kolboom (1986, p. 81)).

Stiegler 2004) of know-how and life skills. In France, the massive strikes of 1912 and 1913 against Taylorism enabled the physiology of work to position itself as a legitimate science (Amar 1914). A few years later, Henry le Chatelier gave Taylorism its academic credentials by lobbying France's highest scientific authorities to promote the merits of the American engineer (Letté 2004).

However, it would be quite wrong to see the influence of Taylorism solely in large-scale industry, as small-scale commerce and administration were soon called upon to adapt to the new forms of work optimization: grammatization and proletarization were part of the anthropological transformation brought about by industrialization, which spared almost no social sphere. While Marxism has long denigrated the employee (suspected of bourgeois connections) as the worker's negative double, historiography shows the extent to which offices and stores were the target of massive proletarization at the dawn of the 20th century in Europe: sociologist Roger Girod (1961), for example, highlights the kinship between the lathe and the accounting machine. Historian Delphine Gardey's work on scriptural materiality reveals how economic exploitation and gender oppression combine effectively in administrative or accounting writing and, more specifically, in typing (Gardey 2001, 2008). In the 19th century, the office worker was generally a well-educated man whose primary skill was writing and who was familiar with – and aspired to climb – the ranks of the company, business or administration where he worked: the office was a masculine space. At the end of the administrative revolution in France in the early 20th century, offices (with the exception of directorates) were predominantly occupied by women with a basic level of education, engaged in routine tasks that were often mechanized, were subject to a strict hierarchy and had no career prospects. This massive transformation, despite taking place within the so-called "tertiary" sector, is archetypal evidence of the structuring elements of the industrialization process.

Since at least the French Revolution, the grammatization of writing has been booming, owing to shorthand, which has infused the legal and business worlds from political assemblies. As Gardey shows, liberal capitalist societies are overwhelmingly societies of the written word, which follows goods in all its forms and in all its transports and needs a certain standardization: just like our digital fonts (which are their legacy), specific graphic forms defined by the shape of the elements, the size of the body of the letters, the proportions of the breaks, the type of line spacing, etc., mark different uses. The latter scripts are called *bâtarde*, *coulée*, *grosse coulée*, *ronde*, *écriture anglaise*, etc. The development of the first typewriters was a technical adaptation to this global scriptural trend: initially simple self-printing machines (letters are selected on a rotating dial), they retain traces of the production lines on which they are integrated, such as the carriage return operated by a pedal, identical to that on sewing machines. Gradually becoming part of administrative culture, they accelerated the formalization of writing and materialized the possibility of optimizing its production. International typing competitions were held until the middle of the 20th century, and a number of celebrities emerged, such as Margaret Hamma operating an IBM Electromatic Typewriter, who, on June 20, 1941, achieved a record of 149 words typed per minute (peaking at 228: more than three words per second!) during the hour-long contest, which was both a testing ground and an advertising showcase for manufacturers (as in motor racing today). Champion performances were dissected by scientists to extract the ultimate method, and machines gradually engramed the ideal actions and postures into their very morphology. These engineers worked to introduce Taylorism into offices, following the example of psychologist Jean-Maurice Lahy (1872–1943) in France, or Henry Leffingwell (1876–1934) in the United States – president of the Taylor Society and member of the American Management Association of Mechanical Engineers, who in 1917 published a method for applying the scientific organization of work to offices and administrations (Leffingwell 1917).

The cost price of a typed line obsessed these new managers, who even imposed the furniture and uniforms of the hundreds of women employed in these secretarial pools, whose working hours, posture and typing techniques were calculated with profitability in mind. While this process of grammatization, proletarization and mechanization has stretched over several decades in France, the various stages sometimes followed one another much more quickly: in 1901, the colonial administration in Dakar, French West Africa, formalized and standardized handwriting, only to impose the switch to typewriters the following year (Gardey 2008).

These dynamics have a profound effect on the social status of office workers. In small businesses, clerks had to open the cash register in the morning and close it in the evening, sometimes resulting in working days of up to sixteen hours (compared with an average of 12 hours in offices), to such an extent that a working week that allowed them to finish on Saturdays at 4 pm became a massive demand. Paternalism and proletarization then mutually train each other, and the union of professional employers' associations claimed to defend these women by exploiting them:

> It is mostly women who do the shorthand and typing; but in these jobs, they have to be present first thing in the morning when the mail is opened, take the notes needed to write the correspondence, then write the mail during the day and bring it in the evening for signing [...]. If the business were obliged to have these women for less time than the rest of the staff, it would be absolutely impracticable, and to

> do without them would be to do away with one of the best jobs women have. (quoted in Gardey (2001, p. 93))

Some of them nevertheless organized themselves within the *Syndicat des femmes sténographes et dactylographes* (Union of Women Stenographers and Typists) to fight against the harmful effects of this proletarization. Whether they were the script hands of big industry, such as Renault after 1911, or working for a US mail-order company in 1912, these women saw the very nature of office work evolve: the grammatization of writing skills was part of a general proletarization of women's office work. Standardizing and mass-replicating optimized techniques consolidate the pressure of the industrialized market.

Fordism extended and continued the Taylorian enterprise, adding the conveyor belt transport of parts around the factory to the grammatization of tasks (Hughes 1989) – the famous assembly line from Chaplin's *Modern Times*. Owing to the economies of scale achieved by standardized mass production (8 million Ford T models were sold in the United States), Ford was able to considerably increase workers' wages (up to 5 dollars a day, compared to 2–3 dollars elsewhere) while simultaneously creating immense outlets, since workers at his factory could now purchase the goods produced there, thus inaugurating the Fordist compromise and hence the grammatization of consumption. Threshold effects were created by massification, and Ford put Taylor's adage into practice: "In the past, the man has been first; in the future, the system must be first" (Taylor 1919, p. 7). The increase in workers' wages also enabled him to combat the high absenteeism generated by the harsh Taylorist working conditions (due to an ingenious bonus system designed to retain the workforce).

Once this strategy was implemented, the grammatization of manual labor would continue to grow: in the context of postwar reconstruction, Toyotism sought to optimize production by reducing inventory costs through just-in-time production and by enlisting bits and pieces of workers' expertise in quality circles: microrewards paid to the most efficient workers were supposed to encourage workers to mobilize their expertise and to give their opinion on production optimization. Today, lean management inherits this long chain of know-how grammatization and goes as far as to formalize manners, or even soft skills – for example, through group activities designed to bind teams together, or conversely, voluntary destabilization designed to encourage employees to do their best, always with a view to productivity (Linhart 2012).

This long history of the proletarization of living know-how, briefly summarized here, potentially concerns all sectors of human activity and even includes some unexpected moments. The First World War was the scene of a military revolution (Goya 2014): the unprecedented power deployed by artillery meant that the former mobile warfare – initially to avoid bloodshed – became a war of position. Henceforth, in the words attributed to the future Marshal Pétain, "artillery conquers and infantry occupies" (Le Gall 2014, p. 198). The fate of battles thus depended more on the supply of ammunition than on the strategy deployed. The artillerymen themselves, because of the immense destructive radius of shells, were mostly called upon to relentlessly operate the machines of war rather than to fine-tune them within the framework of a constructed plan. Gunners were no longer skilled marksmen, as they had been in the Royal Navy of the 17th century, but simple workers producing industrialized violence. While many factors contributed to the dehumanization of trench warfare, the absurdity of a death committed anonymously miles away by proletarianized men, whose actions made no sense, certainly contributed.

Nevertheless, the machine has no real technical skills, so it needs the help of human beings to temper randomness and smooth out the irregularities of reality. The carpenter Arthur Lochmann describes the technical act of sawing as a permanent mix of intuition, incorporation and reinvention induced by the necessary adaptation to the infinite and multiple variations of reality (Lochmann 2019); the machine is, by essence, repetitive and therefore imposes a reconfiguration of the world to coincide with its own mechanisms. Mechanical action then overflows, transforms and changes the nature of the operation performed to such an extent that men and women have to invent new techniques and cultures to deal with it. At its limit, the human being is reconfigured to match the needs of the machine. Protesting against the naïve evolutionary chronology that imagines the history of techniques as a linear progression from simple to complicated, anthropologist Tim Ingold rejects the assimilation of technical know-how to mere execution: the craftsman's actions are guided by a keen perception of the activity in progress, whereas the machine's movements are predetermined. For Ingold (2011), the transition from tool to machine therefore primarily implies a "displacement of the producer, as a person, from the center to the periphery of the production process. In other words, we are dealing with a history, an externalization, not a complexification" (pp. 289–290).

The introduction of steam-powered machines into the French countryside in the 19th century was a laboratory for agricultural machinery, at a time when rural areas seemed untouched by the transformations affecting large cities and industrial work. The use of steam threshing machines requires the reinvention of peasant farming to adapt workers to the machine.

> The steam engine, complex and dangerous, freezes workstations, regulating their intensity and often requiring a "standard" worker

> capable of keeping up the pace, leading to the exclusion of the elderly, women, young or disabled. Whether for installation, heating, feeding the boiler, greasing the gears or regular cleaning, the use of threshing machines required new skills. [...] By the end of the 19th century, it was increasingly rare for peasants to own their own machines; it was now the threshing contractor who owned them, maintained them and moved them from farm to farm, usually paid by the hour. (Jarrige 2019, pp. 56–57)

The mechanical timeline takes precedence and reconditions human activity: *we were using tools, and we are now used by machines*. The introduction of automatic gates in the Paris metro is a good illustration of the consequences of mechanization. Punch presses – which became punching machines after the First World War – marked the stations with their presence, contributing to passenger safety and information (Dumont 1998): "There was no barrier, but contact, exchange – even of a ticket – with someone" (Racle 1983, p. 111). Where the punching tool was an extension of the physiological and social body, this validator integrated into the automatic terminal now includes the human in its extension: the user who passes through the turnstile must conform his rhythm to that of the machine, which is unable to adapt to the diversity of situations – hence the existence of additional doors that can be opened manually on the side for passengers outside the norm (wheelchairs, bulky suitcases, baby carriages, etc.). In other words, the punching machine was acted upon and gave rise to a coupling between technique and humanity; automated gates are subjected to and have generated a spatiotemporal dissociation between the mechanical and the human. The technological burst of industrial mechanics has created a disruption in the anthropotechnical world.

"All labor, whenever it is placed in competition with a slave, whether the slave be human or mechanical, must accept the slave's conditions" – a letter reproduced in Noble (2016, pp. 223–227). Norbert Wiener, mathematician and founder of cybernetics, wrote to Walter Reuther, president of the American United Automobile Workers union, in August 1949. Wiener had already grasped what recent historical studies of industrialization have shown: mechanization is, above all, a transformation of social relations, and the machine "substitutes itself not for the energy of the workers, but for their judgment" (Noble 2016). The machines introduced in recent centuries have helped crystallize power relations based on the regimentation and exploitation of man by man.

To grammatize know-how is to formalize sedimented time, to leave behind the living world of permanent, controlled human reinvention in favor of forced planning. Like the Bourdieusian *habitus*, know-how, as a projection of incorporated sociotechnical structures, tends to reproduce the conditions of its production and to produce the conditions of its reproduction (Bourdieu and Passeron 1970): it is a reinvestment of that which is known in the uncertain world of materiality and, as such, a permanent invention in a constrained situation. Proletarianizing expertise means breaking the subtle interplay between human beings and technique and making techno-anthropological transduction bend under industrial economic constraint.

6.5. References

Amar, J. (1914). *Le moteur humain et les bases scientifiques du travail professionnel*. Dunod et Pinat, Paris.

Amiech, M. (2003). Entre personnalisation et gestion standardisée : les paradoxes des centres d'appels. In *Du politique dans les organisations. Sociologie des dispositifs de gestion,* Broussard, V. and Maugeri, S. (eds). L'Harmattan, Paris.

Auroux, S. (1994). *La révolution technologique de la grammatisation*. Mardaga, Liège.

Baptiste, F., Charre, P., Ganne, B. (1988). *Mégissiers, tanneurs et papetiers à Annonay au début du siècle. Approche historique de l'espace urbain et de l'espace social*. Groupe lyonnais de sociologie industrielle, Bron.

Barron, G. (2022). "Avec les moyens du bord" : la réparation à bord des navires français, XVIIIe-XIXe siècles. In *Les réparations dans l'histoire. Cultures techniques et savoir-faire dans la longue durée*, Bernasconi, G., Carnino, G., Hilaire-Pérez, L., Raveux, O. (eds). Presses des Mines, Paris.

Baudeau, N. (1910). *Première introduction à la philosophie économique, ou analyse des états policés*. Geuthner, Paris [Original published 1771].

Beaud, C. (1995). Industrie. In *Dictionnaire du Second Empire*, Tulard, J. (ed.). Fayard, Paris.

Benoît, P. (1998). Au four et au moulin : innovation et conjoncture. In *L'innovation technique au Moyen Âge. Actes du VIe congrès international d'archéologie médiévale*, Beck, P. (ed.). Errance, Paris.

Benoît, S. (2020). *D'eau et de feu : forges et énergie hydraulique, XVIIIe-XXe siècle. Une histoire singulière de l'industrialisation française*. PUR, Rennes.

Benvegnù, C. and Gaborieau, D. (2017). Produire le flux. L'entrepôt comme prolongement d'un monde industriel sous une forme logistique. *Savoir/Agir*, 39(1), 66–72.

Berg, M. (1980). *The Machinery Question and the Making of Political Economy: 1815-1848*. Cambridge University Press, Cambridge.

Berg, M. (1985). *The Age of Manufactures: 1700-1820*. Fontana, London.

Berg, M. (1993). What difference did women's work make to the industrial revolution? *History Workshop*, 35, 22–44.

Berg, M. (2014). Skill, craft and histories of industrialisation in Europe and Asia. *Transactions of the Royal Historical Society*, 24, 127–148.

Berg, M. and Hudson, P. (1992). Rehabilitating the industrial revolution. *The Economic History Review*, 45(1), 24–50.

Bernardi, P. and Verna, C. (2001). Travail et Moyen Âge : un renouveau historiographique. *Cahiers d'histoire, revue d'histoire critique*, 83, 27–46.

Bernier, S. (2007). Perceptions des risques industriels et nucléaires. Enjeux, négociations et construction sociale des seuils d'acceptation des risques. PhD Thesis, Université François Rabelais, Tours.

Bilbao, Z.E. (2020). Évolutions de la coroplathie archaïque en Grande Grèce, de l'étude de cas à l'analyse régionale. *Quaderni ticinesi – Numismatica e Antichità Classiche*, 49, 11–25.

Blundel, R. and Tregear, A. (2006). From artisans to "factories": The interpenetration of craft and industry in English cheese-making, 1650-1950. *Enterprise and Society*, 7(4), 705–739.

Bourdieu, P. and Passeron, J.-C. (1970). *La reproduction*. Minuit, Paris.

Braverman, H. (1998). *Labor and Monopoly Capital: The Degradation of Work in the Twentieth Century*. Monthly Review Press, New York.

Bruland, K., Gerritsen, A., Hudson, P., Riello, G. (eds) (2020). *Reinventing the Economic History of Industrialisation*. McGill-Queen's University Press, Montreal.

Caliste, L. (2021). Le Lodévois et ses marges. Industries, ateliers et entreprises (vallées de la Lergue et de l'Hérault, XVe siècle). PhD Thesis, Université Paris 8, Paris.

Carnevali, F. (2003). Golden opportunities: Jewelry making in Birmingham between mass production and specialty. *Enterprise and Society*, 4(2), 272–298.

Carnino, G. (2014a). L'académicien et le pêcheur. Entre science, État et industrie : l'invention de la pisciculture en France, 1840-1880. In *Savants et inventeurs en-tre la gloire et l'oubli (actes du 134e congrès national des sociétés historiques et scientifiques – CTHS, 12 avril 2009)*, Bret, P. and Pajonk, G. (eds). CTHS, Paris.

Carnino, G. (2014b). Louis Pasteur. La science pure au service de l'industrie. *Le Mouvement Social*, 248(3), 9–26.

Carnino, G. (2015). *L'invention de la science. La nouvelle religion de l'âge industriel*. Le Seuil, Paris.

Carnino, G. and Marquet, C. (2018). Les datacenters enfoncent le cloud : enjeux politiques et impacts environnementaux d'internet. *Zilsel*, 3, 19–62.

Carnino, G. and Marquet, C. (2020). Du mythe de l'automatisation au savoirfaire des petites mains : une histoire des datacenters par la panne. *Artefact. Techniques, histoire et sciences humaines*, 11, 161–188.

Carnino, G. and Marquet, C. (2022). Cooling, quick fix et spaghetti cloud dans l'univers du datacenter. Changement d'échelle et industrialisation du numérique. *Artefact. Techniques, histoire et sciences humaines*, 17, 309–335.

Caron, F. (1997). *Les deux révolutions industrielles du XXe siècle*. Albin Michel, Paris.

Caspard, P. (1979). *La Fabrique-Neuve de Cortaillod*. Publications de la Sorbonne, Paris.

Cerman, M. and Ogilvie, S. (eds) (1994). *Protoindustrialisierung in Europa*. Verlag für Gesellschaftskritik, Vienne.

Chessel, M.-E. (2012). *Histoire de la consommation*. La Découverte, Paris.

Ciarla, R. and de Luca, A. (2005). *L'armée éternelle. Les soldats du premier empereur de Chine*. National Geographic, Paris.

Conner, C.D. (2011). *Histoire populaire des sciences*. L'échappée, Paris.

Coriat, B. (1982). *L'atelier et le chronomètre. Essai sur le taylorisme, le fordisme et la production de masse*. Christian Bourgois, Paris.

Cunningham, A. and Williams P. (1993) De-centring the "Big Picture": "The Origins of Modern Science" and the modern origins of science. *The British Journal for the History of Science*, 26–4, 407–432.

Daviet, J.-P. (1997). *La société industrielle en France. 1814-1914*. Le Seuil, Paris.

Descartes, R. (1992). *Méditations métaphysiques*. Flammarion, Paris [Original published 1641].

Dessert, D. (1996). *La Royale. Vaisseaux et marins du Roi-Soleil*. Fayard, Paris.

Dewerpe, A. (1985). *L'industrie aux champs. Essai sur la proto-industrialisation en Italie du nord (1800-1880)*. École Française de Rome, Rome.

Donzé, P.-Y. (2011). The hybrid production system and the birth of the Japanese specialized industry: Watch production at Hattori & Co. (1900-1960). *Enterprise and Society*, 12(2), 356–397.

Dumont, G.-F. (1998). *Transports urbains et espace économique. Nouveaux espaces et systèmes urbains*. Sedes, Paris.

Espagnol, S. and Leterme, P. (2010). *Élevages et environnement*. Éducagri, Dijon.

Flacher, D. and Pelletan, J. (2007). Le concept d'industrie et sa mesure : origines, limites et perspectives. Une application à l'étude des mutations industrielles. *Économie et Statistique*, 13–46, 405–406.

Fontaine, P. (1992). Le concept d'industrie au XVIIIe siècle : à la recherche d'une acception oubliée. *Économies et Sociétés, série "Œconomia"*, 3, 7–33.

Fontaine, P. (1993). The concept of industrie from the physiocrats to J.-B. Say. *Contributions to Political Economy*, 12, 89–97.

Fontaine, M. and Vigna, X. (2019). La désindustrialisation, une histoire en cours. *Revue d'histoire*, 144(4), 2–17.

Foucault, M. (1975). *Surveiller et punir*. Gallimard, Paris.

Fressoz, J.-B. (2015). Mundus œconomicus. Révolutionner l'industrie et refaire le monde après 1800. In *Histoire des sciences et des savoirs 2 : modernité et globalisation*, Kapil, R. and Otto, S.H. (eds). Le Seuil, Paris.

Fridenson, P. (1972). Une industrie nouvelle : l'automobile en France jusqu'en 1914. *Revue d'histoire moderne et contemporaine*, 19(4), 557–578.

Gardey, D. (2001). *La dactylographe et l'expéditionnaire. Histoire des employés de bureau (1890-1930)*. Belin, Paris.

Gardey, D. (2008). *Écrire, calculer, classer. Comment une révolution de papier a transformé les sociétés contemporaines (1800-1940)*. La Découverte, Paris.

Gerritsen, A. and Riello, G. (eds) (2016). *The Global Lives of Things. The Material Culture of Connections in the Early Modern World*. Routledge, London/New York.

Giedion, S. (1980). *La mécanisation au pouvoir*. Centre Georges Pompidou, Paris.

Girod, R. (1961). *Études sociologiques sur les couches salariées, ouvriers et employés*. Marcel Rivière, Paris.

Goya, M. (2014). L'armée française et la révolution militaire de la Première Guerre mondiale. *Politique étrangère*, 79(1), 87–99.

Graeber, D. and Wengrow, D. (2021). *The Dawn of Everything: A New History of Humanity*. Penguin, London.

Grand View Research (2019). Surfing equipment market size, share & trends analysis report by product (surfing boards, apparel & accessories). Grand View Research.

Hahn, B. (2020). *Technology in the Industrial Revolution*. Cambridge University Press, Cambridge.

Halleux, R. (2009). *Le savoir de la main. Savants et artisans dans l'Europe pré-industrielle*. Armand Colin, Paris.

Hatzfeld, N. (2022). L'usine comme objet technique ? Un éclairage sur le XXe siècle industriel. *Artefact*, 17, 285–307.

Hauser, H. (1927). *Les débuts du capitalisme*. Alcan, Paris.

Hayami, A. (1986). A great transformation. Social and economic change in sixteenth and seventeenth-century Japan. *Bonner Zeitschrift für Japanologie*, 8, 3–13.

Herment, L. (2017). Y a-t-il eu une révolution industrieuse dans le Bassin parisien au XIXe siècle ? *Revue d'histoire moderne et contemporaine*, 64(4), 105–133.

Herrigel, G. and Zeitlin, J. (2010). Alternatives to varieties of capitalism. *The Business History Review*, 84(4), 667–674.

Hilaire-Pérez, L. (1996). *L'expérience de la mer. Les Européens et les espaces maritimes au XVIIIe siècle*. Seli Arslan, Paris.

Hilaire-Pérez, L. (2000). *L'invention technique au siècle des Lumières*. Albin Michel, Paris.

Hilaire-Pérez, L. (2013). *La pièce et le Geste. Artisans, marchands et savoir technique à Londres au XVIIIe siècle*. Albin Michel, Paris.

Hilaire-Pérez, L. and Jarrige, F. (2021). La machine à vapeur démythifiée. *Les Collections de l'Histoire*, 91, 24–31.

Hinrichs, P. and Kolboom, I. (eds) (1986). *Absolut modern sein: Zwischen Fahrrad und Fliessband. Culture technique in Frankreich, 1889-1937*. NGBK/Elefanten Press Vlg, Berlin.

Hughes, T.P. (1989). *American Genesis. A Century of Invention and Technological Enthusiasm, 1870-1970*. Penguin, New York.

Humphries, J. (2003). Child labor: Lessons from the historical experience of today's industrial economies. *The World Bank Economic Review*, 17(2), 175–196.

Ingold, T. (2011). L'outil, l'esprit et la machine. Une excursion dans la philosophie de la "technologie". *Techniques and Culture*, 54(55), 288–311.

Inskter, I. and O'Brien, P. (eds) (2004). *History of Technology, Volume 25: The Global History of the Steam Engine*. Bloomsbury, London.

Jarrige, F. (2014). *Technocritiques. Du refus des machines à la contestation des techno-sciences*. La Découverte, Paris.

Jarrige, F. (2019). Batteuses à manège ou batteuses à vapeur ? *Artefact*, 11, 39–65.

Jarrige, F. (2022). Origine et signification du mot "industrie". Retour sur une querelle sémantique de l'entre-deux-guerres. *Artefact*, 17, 265–283.

Jarrige, F. (2023). *La ronde des bêtes. Le moteur animal et la fabrique de la modernité*. La Découverte, Paris.

Judet, P. (2019). *La nébuleuse métallurgique alpine (Savoie-Dauphiné, fin XVIIIe-fin XIXe siècle). Apogée, déclin et éclatement d'un territoire industriel*. Presses universitaires de Grenoble, Grenoble.

Kadish, A. (1986). *Apostle Arnold: The Life and Death of Arnold Toynbee, 1852-1883*. Duke University Press, Durham.

Le Gall, E. (2014). *Une entrée en guerre. Le 47e régiment d'infanterie de Saint-Malo au combat (août 1914-juillet 1915)*. Codex, Talmont-Saint-Hilaire.

Ledu, S. and Germain, C. (2016). *La révolution industrielle*. Milan, Toulouse.

Leffingwell, H. (1917). *Scientific Office Management. A Report of the Application of the Taylor System of Scientific Management to Offices*. A. Shaw Company, Chicago.

Lemarié, J. (2018). *Surf : une histoire de la glisse, de la première vague aux Beach Boys*. Arkhê, Paris.

Letté, M. (2004). *Henry Le Chatelier (1850-1936) ou la science appliquée à l'industrie*. PUR, Rennes.

Linhart, D. (2012). Quand le management se fait Dibbouk. *Nouvelle revue de psychosociologie*, 13(1), 31–42.

Lochmann, A. (2019). *La vie solide. La charpente comme éthique du faire*. Payot, Paris.

Malm, A. (2016). *Fossil Capital. The Rise of Steam Power and the Roots of Global Warming*. Verso, Londres.

Marx, K. (1993). *Le Capital, volume I*. PUF, Paris [Original published 1867].

Millet, G. (2012). Les tanneries du XVIIIe au XXe siècle [Online]. Available at: http://ccha.fr/wp-content/uploads/2012/01/Genevi%C3%A8ve-MilletLes-Tanneries-du-18%C3%A8me-au-20%C3%A8me-si%C3%A8cles.pdf [Accessed 14 April 2022].

Mokyr, J. (1990). *The Lever of Riches. Technological Creativity and Economic Progress*. Oxford University Press, New York.

Moreno, C.B. (2017). Révolution industrieuse ou simple survie ? Stratégies familiales dans une zone protoindustrielle de Catalogne au XVIIIe siècle. *Revue d'histoire moderne et contemporaine*, 64(4), 59–83.

Mumford, L. (1974). *Le mythe de la machine*. Fayard, Paris.

Navel, G. (1979). *Travaux*. Gallimard, Paris.

Noble, D. (2016). *Le progrès sans le peuple*. Agone, Marseille.

O'Brien, P. (2012). *Two Paths to the 20th Century: Economic Growth in Britain and France 1789-1914*. Routledge, London.

Pack, S.J. (1991). *Capitalism as a Moral System: Adam Smith's Critique of the Free Market Economy*. Edward Elgar Publishing, Aldershot.

Pagès, G. and Verna, C. (2022). L'invention de l'industrie antique et médiévale. *Artefact*, 17, 243–264.

Paul, H. (1985). *From Knowledge to Power: The Rise of the Science Empire in France, 1860-1939*. Cambridge University Press, Cambridge.

Perrot, M. (1983). De la manufacture à l'usine en miettes. *Le Mouvement Social*, 125, 3–12.

Pestre, D. (2003). *Science, argent et politique. Un essai d'interprétation*. INRA, Paris.

Pfister, U. (1992). *Die Zürcher Fabriques, Protoindustrielles Wachstum vom 16. zum. 18. Jahrhundert*. Chronos, Zürich.

Polanyi, K. (1983). *La Grande Transformation. Aux origines politiques et économiques de notre temps*. Gallimard, Paris.

Racle, G. (1983). Les langages du métro. *Communication and Langages*, 57, 107–114.

Raggi, P. (2019). *La désindustrialisation de la Lorraine du fer*. Classiques Garnier, Paris.

Rauhut, C. (2018). The non-simultaneous as a research tool for late nineteenth-century construction sites. In *Producing Non-Simultaneity. Construction Sites as Places of Progressiveness and Continuity*, Eike-Christian, H. and Christoph, R. (eds). Routledge, London/New York.

Rediker, M. (2010). *Les forçats de la mer marins, marchands et pirates dans le monde anglo-américain, 1700-1750*. Libertalia, Montreuil.

Riello, G. and Parthasarathi, P. (eds) (2009). *The Spinning World: A Global History of Cotton Textiles, 1200-1850*. Oxford University Press, Oxford.

Rioux, J.-P. (1971). *La révolution industrielle, 1780-1880*. Le Seuil, Paris.

Sabel, C. and Zeitlin, J. (1985). Historical alternatives to mass production. Politics, markets and technology in nineteenth-century industrialization. *Past and Present*, 108, 133–176.

Sabel, C. and Zeitlin, J. (eds) (1997). *World of Possibilities. Flexibility and Mass Production in Western Industrialization*. Cambridge University Press, Cambridge.

Scranton, P. (1997). *Endless Novelty: Specialty Production and American Industrialization, 1865-1925*. Princeton University Press, Princeton.

Singleton, M. (2007). Croyance et décroissance. *Entropia*, 3, 37–44.

Smith, A. (1976). *An Inquiry into the Nature and Causes of the Wealth of Nations*. Oxford University Press, Oxford [Original published 1776].

Stiegler, B. (2004). *Mécréance et discrédit, vol. 1 : la décadence des démocraties industrielles*. Galilée, Paris.

Taylor, F.W. (1919). *The Principles of Scientific Management*. Harper and Brothers, New York/London.

Thébert, Y. (2000). Transport à grande distance et magasinage de briques dans l'Empire romain. Quelques remarques sur les relations entre production et consummation. In *Brique antique et médiévale. Production et commercialisation d'un matériau. Actes du colloque international : Saint-Cloud, 16-18 novembre 1995*, Boucheron, P., Broise, H., Thébert, Y. (eds). École française de Rome/Boccard, Rome/Paris.

Thompson, E.P. (2012). *La formation de la classe ouvrière anglaise*. Le Seuil, Paris.

Tsing, A.L. (2012). On nonscalability. The living world is not amenable to precision-nested scales. *Common Knowledge*, 18(3), 505–524.

Verley, P. (1997). *L'échelle du monde. Essai sur l'industrialisation de l'Occident*. Gallimard, Paris.

Verna, C. (2017). *L'industrie au village. Essai de micro-histoire (Arles-sur-Tech, XIVe et XVe siècles)*. Les Belles Lettres, Paris.

Versieren, J. and De Smet, B. (2019). The strategies of fear, the commercialization of society, and the rise of the factory system in the low countries during the eighteenth and nineteenth centuries. In *Historicizing Fear. Ignorance, Vilification, and Othering*, Boyce, T.D. and Chunnu, W.M. (eds). University Press of Colorado, Denver.

Vincent, J. (2021). Marx, Toynbee et la "révolution industrielle". *Les Collections de l'Histoire*, 91, 32–35.

de Vries, J. (2008). *The Industrious Revolution. Consumer Behavior and the Household Economy, 1650 to the Present*. Cambridge University Press, Cambridge.

Wichterich, C. (1999). *La femme mondialisée*. Solin/Actes Sud, Arles.

7

From Standards to Commodification (and Back Again): Histories of Technology and Capitalism[1]

Joris MERCELIS[1,2]

[1] *Heinrich Heine University Düsseldorf, Germany*

[2] *Johns Hopkins University, Baltimore, USA*

7.1. Introduction

Does technology matter to capitalism, and vice versa? One might think that the answer to both of these questions is a resounding “yes!” The importance of technology, after all, was already stressed by classical theorists of capitalism ranging from “innovation prophet” Joseph Schumpeter to Karl Marx, whose views on the relationship between technical and socioeconomic change were much more sophisticated than what might be suggested by his famous dictum that “the hand-mill gives you society with the feudal lord; the steam-mill, society with the industrial capitalist” (MacKenzie 1984; McCraw 2007). The rhetoric of successive industrial “revolutions”, which emerged in the second half of the 20th century and has been promoted by the World Economic Forum in recent years (Schwab 2017), has been even more technology-centered. The same admittedly does not apply to all of the

1 This chapter partly draws on literature covered in the “Science and the Marketplace” seminar that I taught at Johns Hopkins University in spring 2021 and 2023. I am grateful to the students in this course for many stimulating discussions.

History of Techniques,
coordinated by Guillaume CARNINO and Xavier GUCHET.

many varieties of capitalism that social scientists and historians have identified (see, most recently, Schakenbach Regele (2023)). However, this literature has also highlighted the technical changes associated with transitions from, say, "merchant" and "industrial" capitalism to "postindustrial" capitalism (Kocka 2016). Information and communication technology has been especially generative in terms of stimulating new perspectives on capitalism; Google's and Facebook's targeted online advertising techniques, for example, are at the heart of Shoshana Zuboff's (2019) widely discussed analysis of "surveillance capitalism" (see also Lauer and Lipartito (2021)).

Nevertheless, despite these and many other examples, historians of technology have been relatively slow to respond to the renewed academic and societal interest in capitalism since the end of the Cold War and especially since the turn of the 21st century[2]. According to a recent article in *Technology and Culture*, indeed, technology historians still need to be convinced that "capitalism matters". More specifically, the author of this article, Edmund Russell (2022, p. 32), proposes to "elevate the history of capitalism from a minor theme to a research agenda in the history of technology". He calls for more engagement with the so-called new history of capitalism (NHOC), a largely US-based historiographical movement that has sought to bring together economically oriented research conducted across a range of different historical subdisciplines under the "capitalism" label (Barreyre and Blin 2017; Russell 2022).

Rather than offering a comprehensive review of the intersections between the history of technology, on the one hand, and economic history and the history of capitalism, on the other hand, Russell specifically focuses on finance, using a case study on telegraphy in the mid-19th-century United States to argue for the inseparability of financial and technological innovation. I agree with him that finance has been a relatively marginal topic in the recent historiography of technology – although there are signs that this situation has started to change (MacKenzie 2015; Deringer 2018; Ward 2019)[3] – and an area where important additional work remains to be done. It is also true that, along with slavery and commodification, finance has been a key theme in the NHOC (Hilt 2017,

2 This interest has been especially pronounced in US history departments, where it is associated with the promotion of a "new history of capitalism", but it is also evident elsewhere. See, for example, Kocka and van der Linden (2016). As Kocka explains in that volume, "something like a limited revival" of the capitalism concept began in the 1990s (p. 3).

3 For instance, several papers presented at the 2022 annual meeting of the Society for the History of Technology (SHOT) dealt with finance, including Donald MacKenzie's Da Vinci Medal award address.

pp. 512–513; Levy 2017, pp. 483–484). Moreover, as this chapter illustrates, historians of technology and economic historians have meaningfully engaged with each other's work in multiple other areas, including topics integral to current approaches to the study of capitalism (e.g. the histories of knowledge work and of gendered and racialized divisions of labor in capitalist economies)[4]. By reviewing such literature, it is possible to take stock of what scholars in both subdisciplines can learn and have learned from each other and to identify avenues for future research.

Even in regard to research on a single topic, there have been strikingly different patterns of interaction that, when considered collectively, reveal both the benefits and limitations of subdisciplinary cross-fertilization. During the last three decades, both economic historians and historians of technology and science have been studying the role of "useful and reliable knowledge" – a phrase proposed by Ian Inkster (2006) – in global processes of technoeconomic divergence and convergence. As Dagmar Schäfer and Simona Valeriani (2021) summarized, they initially approached this problem within their own scholarly communities without addressing the findings and methods of researchers in the other subdiscipline. In a second phase, however, Joel Mokyr's *Gifts of Athena* (2002) and several related publications "generated debate that brought historians of science and technology together with economic historians to engage in a new way of looking at the contributions of science and technology to the industrial development from the early modern period to the twentieth century" (Berg 2007, p. 123; Schäfer and Valeriani 2021, p. 329). In many ways, this debate was productive, not least in stimulating additional research on the connections between knowledge and industrial development in previously underexplored non-Western contexts. However, in a third research phase that started around 2010 and continues into the present, historians of technology and science have increasingly challenged the tendency in the global economic history literature to approach the "utility" of knowledge through Eurocentric frameworks in capitalist, growth-enhancing terms (Schäfer and Valeriani 2021, pp. 329–332). Instead, these scholars have begun to culturally and conceptually expand the scope of historical investigations of the usefulness and reliability of knowledge into a wide range of non-Western regions and languages and less frequently take developments in Europe as their reference point (Schäfer and Valeriani 2021). In this respect, in other words, the two subdisciplines appear to be on diverging research paths.

4 The history of knowledge work is a central theme in a recent collection of historical articles on science and capitalism, which engages with NHOC and includes contributions by several historians of technology. See Rieppel et al. (2018). On gendered labor divisions, see, for example, Cowan (1983) and de la Cruz-Fernández (2021). On slavery, technology and capitalism, see, for example, Rood (2017).

What about other problems of (potential) mutual interest? Identifying additional areas where fruitful interactions have taken place for at least limited periods of time is not difficult. Economists and economic historians such as Nathan Rosenberg (1982, pp.19–23, 131–140), for example, questioned the relative importance of studies on "great inventors" and highlighted the significance of technology "diffusion" and maintenance work – themes that have been of great interest to technology historians in recent years[5]. The emergence of intellectual property as a popular research theme in both subdisciplines also involved substantial cross-community knowledge exchanges (for example, Cooper (1991)).

It would be well beyond the scope of this chapter to systematically review the substantial intersections between economic history and the history of technology regarding these and various additional themes over the span of multiple decades. My aims are, therefore, much more modest. Just as Schäfer and Valeriani (2021) approached prior research on the history of "useful and reliable knowledge", I concentrate on one particular topic – in my case, the history of technical standards and standardization processes – to explore whether and how economic and business historians have been able to find common ground with historians of technology and science and how interactions between these scholarly communities might continue in the future. My analysis highlights that the history of standard setting is an area where historians of these different fields have already contributed significantly to the study of capitalism, including to the interpretation of commodification processes of central interest to the NHOC. Moreover, I pay attention to relatively understudied aspects of the relationships and entanglements between commodification and standardization, which would benefit from further research.

7.2. Standards and standardization: economic and technological histories

As with the study of capitalism, research on standards and standardization has been a truly multidisciplinary undertaking involving not only economists and historians but also (among others) sociologists, political scientists and scholars in science and technology studies[6]. The specific questions that researchers have addressed naturally reflect their own disciplinary backgrounds and training. At the

5 Rosenberg was awarded SHOT's prestigious Da Vinci Medal in 1996. On the recent interest in maintenance work, see, for example, Russell and Vinsel (2018).

6 Useful literature reviews emphasizing the multidisciplinary nature of the topic include Moguen-Toursel (2008), Ordóñez and Sánchez (2020), Russell (2005), Slaton and Abbate (2001), Sumner and Gooday (2008), and Timmermans and Epstein (2010).

same time, as Andrew Russell (2005) has argued, "a comprehensive approach to standardization should integrate perspectives" from several different realms, including technology, labor, economics and business. How have historians of technology and economic and business historians taken up this challenge, either independently or in collaboration with each other?

Importantly, the role and importance of technical standards and standardization processes is a theme addressed in various classical works at the intersection of economic and technological history, ranging from David A. Hounshell's (1984) study of the origins of mass manufacturing in the United States and David F. Noble's (1977, Chapter 5) account of the rise of the science-related industrial corporation and the professional engineer to Paul A. David's (1985) attempt to explain the seemingly odd dominance of the QWERTY keyboard. Moreover, standards and standardization are topics that continue to attract interest among both historians of science and technology and economic and business historians, as recent collections of articles in the *Business History Review* and the *Journal of History of Science and Technology* illustrate (Ordóñez and Sánchez 2020; Yates and Murphy 2022); see also Scranton and Fridenson (2013, pp.160–65). Among other things, the more recent literature has cast valuable light on the range of things and experiences that have at least to some extent been standardized, from music listening habits and life design techniques to wheelchairs and chemically defined food (Woods and Watson 2004; Spiekermann 2011; Hui 2012; Meister 2020). In addition, it helps better understand under what conditions standardization processes did and did not lead to reduced variability and what areas of technological and economic life have defied standardization altogether (e.g. Singerman (2018); see also Scranton (1997)).

The extensive resulting body of literature on technical standards and standardization is heterogeneous in many respects, and, as with the "new history of capitalism", scholars in this field do not agree on basic definitions[7]. Nevertheless, useful distinctions between different types of standards and forms of standardization can and have been made. The metric system and other *measurement* standards, for example, are very different from *compatibility* standards, such as those for phone chargers. Process- or result-oriented *performance* standards (e.g. for more environmentally-friendly buildings) can be added as a third category (Russell 2014, pp. 17–18). Additionally, so-called *voluntary consensus* standards negotiated within institutions such as the International Organization for Standardization (ISO) can be contrasted with legally mandated (*de jure*) standards and with *de facto* standards resulting from the widespread adoption of particular products and techniques (Murphy and Yates 2009; Russell 2014, pp. 18–19).

7 On NHOC's approach to defining capitalism, see, for example, Beckert and Desan (2018).

Methodologically, the use of models and econometric methods is one feature that distinguishes some economic history research from studies by historians of technology, science and business[8]. While such methodological differences have doubtlessly created communication barriers of the type common in interdisciplinary research, they have not prevented knowledge exchanges altogether. Methodological choices, indeed, may have been less divisive than exclusively economic research *perspectives*. In their recent history of international voluntary standard-setting, for example, JoAnne Yates (a historian of information technology and management) and Craig N. Murphy (a political scientist) stress the shortcomings of "accounts based purely on economic interest". As an alternative, Yates and Murphy (2021, p. 13) propose to "approach private standardization [...] as an entirely different realm with a very different logic from either commerce or politics, something that developed in response to the greater social complexity that accompanied the pressure toward greater economic integration of industrial capitalism". At the same time, Yates and Murphy recognize the value of economic simulations and models of standard-setting activities and, conversely, some economists and economic historians who have used such tools likewise draw on qualitative historical research (Puffert 2009; Spencer and Temple 2016). Some of the most influential economic analyses of standards (e.g. on the dominance of the QWERTY/AZERTY keyboard layout) aimed to demonstrate to economists that the best technical solutions are not necessarily those that are adopted and that historical research is necessary to uncover important path dependencies (David 1985, 1997).

Although scholars of standards do not necessarily agree on *how* history and economics matter, including with respect to the QWERTY typewriter keyboard (Gardey 1999; Reinstaller and Hölzl 2009), the fact that there has been widespread agreement on the need for historical research approaches among researchers in this field helps explain why economists, economic historians, and historians of technology and business have been able to find substantial common ground. One notable area of agreement concerns the importance of *trust* in the formulation and implementation of standards. At first glance, trust might seem less pertinent than, for example, power hierarchies and considerations of efficiency[9]. Standardization, however, is not just about the voluntary or imposed reduction of variability; rather, as Jean-Paul Gaudillière and Ilana Löwy (1998, p. 10) have observed in relation to

8 For examples of historical studies that use models, see Spencer and Temple (2016) and Puffert (2009). For reviews of models used in the economic literature on standards, see David and Greenstein (1990) and Stango (2004).

9 For an example of a study that uses each of these concepts, see Russell (2008). Power has been central to sociological interpretations of standards (see, for example, Busch (2011, p. 28)).

laboratory research, standardization involves reliability judgments about not only goods but also people and is therefore ultimately "about trust in persons and their doing". Historians of technology and science have explored this theme in comprehensive detail with regard to the history of electrical measurement. One strand of literature has presented standard setting as a crucial activity allowing for "the domination of this world" by aligning local laboratory knowledge and values with imperial networks and agendas (Schaffer 1992, p. 23); see also Hunt (1994, pp. 56–57, 61). Other scholars, however, have stressed that the proclamation of supposedly universal units of measurement (e.g. the ohm as a unit of resistance) did not eliminate the need for trust among 19th-century electrical practitioners. Such standards are also not always necessary to ensure trust (Gooday 2004, pp. 9–16). In the context of laboratory research and measurement, however, networks of trust are central to the development and implementation of standards.

Outside of scientific and engineering laboratories, specific institutional arrangements helped foster trust among the heterogeneous range of actors involved in standard-setting activities. One significant feature of the late 19th- and 20th-century history of standard setting in Europe and North America is that, in the absence of threats to safety and public health, governments have generally been reluctant to take the lead in defining and imposing standards – a reluctance that helps explain why voluntary consensus standards have been so influential[10]. However, in collaboration with industrial and scientific stakeholders, governments did support the creation of public metrology and standardization institutes such as the Physikalisch-Technische Reichsanstalt in Germany (founded in 1887), the British National Physical Laboratory (1899) and the National Bureau of Standards in the United States (1901) (Cochrane 1966; Mosely 1978; Cahan 1989). They also played crucial roles in fostering collaboration between various interest groups and scientific and technical experts to determine standards. In Imperial and Weimar Germany, for instance, the state supported the trust-enabling development of a shared conceptual framework – a chemically defined "nutrient paradigm" – that made it easier for scientists and representatives of the food industry to develop food quality standards (Spiekermann 2018, p. 283). Similarly, in the mid-20th-century United States, public institutions and committees led the way in defining standards and criteria meant to facilitate water quality management. In particular, a sanitary engineer working for the State of California's Water Pollution Control Board, Jack E. McKee, formulated water quality criteria in a publication that became a key reference work for other pollution experts (Hearty 2023, Chapter 4).

10 On the limitations of government-imposed technical standards (e.g. their perceived lack of flexibility), and governments' motives for setting standards in the first place, see, for example, Lundgreen (1997) and Frohlich (2022).

The role of *engineers* in standard-setting processes in capitalist economies and societies is, in fact, another key theme in the works of both historians of technology and economic historians. For example, in his still influential *America by Design: Science, Technology, and the Rise of Corporate Capitalism*, David Noble (1977) discussed the engineer-led standardization movement in the late 19th- and early 20th-century United States in light of his book's central claim that "the history of modern technology in America is of a piece with that of the rise of corporate capitalism" (pp. xxii–xxiii). In Noble's interpretation, the creation and promotion of standards was one of various ways in which scientists, especially engineers, helped stabilize a disorderly capitalist market economy, thus contributing to the "survival of capitalist relations" during the 20th century (p. xxiii). While Noble's portrayal of US engineers as agents of capitalism lacks nuance (e.g. Carlson (1988); Lécuyer (1998)), other studies have confirmed the stabilizing effects of standards and standards institutions on the late 19th- and early 20th-century US economy. Railroad engineers and their employers, for instance, set standards that supported collaborative, controlled forms of innovations (Usselman 2002). More generally, American engineers' obsession with standards was part of a broader effort to establish a more organized society (Russell 2014, p. 27) (see also Galambos (1983)), and even the supposedly hands-off National Bureau of Standards actively intervened to shape markets in the 1920s (Vinsel 2016). In more recent decades too, standards have proven to be relatively effective tools for addressing market failures, as research on automobile safety and emissions performance standards has illustrated (Lee et al. 2010; Vinsel 2019).

During the period studied by Noble (1977), a wide range of engineering societies on both sides of the Atlantic took the lead in developing a new, consensus-based approach to standardization that relied on neither market transactions nor corporate hierarchies (Russell 2014, Chapters 2 and 3); see also Lamoreaux et al. (2003). In the words of one US standards engineer, this movement was "evolutionary and not revolutionary"; it certainly did not radically challenge the power of the large industrial corporations that had started to employ tens of thousands of engineers (Paul Agnew, as cited in Russell 2014, p. 60). Indeed, the president of the Society of Automotive Engineers, Howard E. Coffin, was not exceptional in presenting the development of intercompany standards as a unique and pressing opportunity for engineers to support industry (Thompson 1954, p. 6). Nevertheless, the agendas of standard-setting engineers did not overlap entirely with those of their employers; engineers such as Charles Le Maistre, the secretary of the International Electrotechnical Commission, played more autonomous and indeed even entrepreneurial roles (Yates and Murphy 2021)[11]. Moreover, despite (or owing to)

11 On "standardization entrepreneurs", see also, for example, Graham (2008).

engineers' conservativism, their institutional innovations could have a remarkable long-term impact. Andrew Russell (2014, Chapter 2) even identified their early 20th-century experiments as important roots of the internet-age open standards movement. This long-term impact also applies to actual standards that engineers have developed and the techno-economic trajectories that these standards helped set off. George Stephenson's rail gauge of 4 feet, 8 ½ inches (1435 mm), for instance, famously initiated a new epoch in the history of track gauges (Puffert 2009). The convenience of inch-based standards such as the screw thread standard of William Sellers, a 19th-century Philadelphian machine tool manufacturer and mechanical engineer, contributed to the successful, consequential opposition of US engineers and industrialists to the introduction of the metric system in their country (Sinclair 1969; Mihm 2022).

Given the broad significance of the technical standards created by engineers and other experts, it is important to note that the socioeconomic impact of standards has been decidedly unequal. Historians of technology and labor have rightly emphasized this point. For example, Amy Slaton (2001, p. 65) argued that the quality control standards and specifications for concrete issued by professional associations such as the American Society for Testing and Materials were deliberately ambiguous to ensure that the judgment of college-educated engineers was needed to complete tests and inspections at construction sites. Accordingly, perhaps counterintuitively, the introduction of standards by no means led to a (full) standardization of construction work. Rather, such standards and specifications reinforced already existing social hierarchies and, like other cases, including the history of the Internet's transmission control protocol, can illustrate how "standardization can sometimes be a means to the end of redistributing work, knowledge, and authority among producers or between producers and the users of their products" (Slaton and Abbate 2001, p. 124).

Sociologist Lawrence Busch (2011, Chapter 1) has likewise stressed that standard setting is ultimately about *power* and proposes a broad perspective on standards as "instruments that wield the power to create realities that serve purposes" (Busch and White 2012, p. 246). Economists and legal scholars, by comparison, have specifically explored the relationships between technical compatibility standards and *market power* (e.g. David and Greenstein (1990, pp. 21–24)). Historical research on this question has shown how, in certain circumstances, (intercompany) standards were especially valuable to small manufacturers, for instance, to guarantee them more reliable supplies of parts (Thompson 1954); see also Vermij (2010, p. 244). In other cases, however, oligopolistic or (quasi-)monopolistic firms were able to exploit standards to entrench their dominant market shares. In the mid-20th-century photographic industry, for example, the Eastman Kodak Company leveraged de facto compatibility standards resulting from its near-monopoly over film production to strengthen its position in related markets, including camera manufacturing and photofinishing (Brock 1983; Kadiyali 1998). Either independently or as part of industrial consortia, multinational enterprises such as RCA, Microsoft and Philips have engaged in extensive "standard wars", the outcomes of which have shaped the structure and organization of entire industries and determined the survival (or not) of individual companies and technologies[12]. In areas such as computing and information technology, capitalist processes of innovation and monopoly formation cannot be fully understood without considering standards and standardization.

7.3. Standardization and commodification

Existing economic and technological histories of standard-setting have much to offer historians of capitalism, whether the topic is the role of technical experts in economic and political-economic decision-making or the importance of network externalities in the digital age. More work remains to be done, however, and future research on standards could more directly address ongoing research agendas in the history of capitalism. In this final section of this chapter, I maintain that the relationships and entanglements between standardization and commodification processes represent one promising avenue for additional research. Claims that everything from nature to love is being turned into a commodity – a product that can be bought and sold – are central to contemporary critiques of capitalism (Hall 2023). Both historians and social scientists have described commodification and standardization as closely related phenomena, in areas ranging from healthcare to trade not only in goods such as wheat, sugar and coffee, but also in enslaved people (Cronon 1991, p. 132, pp. 145–47; Timmermans and Almeling 2009; Rosenthal 2018, pp. 137-43; Gago 2022, pp. 238-9, p. 245). However, the historical literature rarely goes beyond the important insight that standardization can enhance commodification by facilitating the sale of interchangeable products over long distances. What about the alternative view that "commodification may lead to standardization" (Ye et al. 2018, p. 51)? In what follows, I suggest that research at the intersection of economic and technological history can offer valuable insights into the conditions under which standardization and commodification have served as stepping stones for each other while also highlighting the mutual shaping of commodification and standardization processes. Finally, I point to the role of standards as (potential) enablers of *decommodification*, that is, the partial or complete abandoning of market logics in a particular social sphere.

12 On the role of industrial consortia in standard-setting processes, see Yates and Murphy (2021, pp. 241–331). On standards battles, see Shapiro and Varian (1999).

The available historical evidence points to significant connections and interdependencies between commodification and standardization. One of the strongest arguments in favor of treating standardization as a prerequisite for commodification is that, in the words of Lawrence Busch (2011, p. 81), "a money society is virtually impossible" without technical standards governing monetary exchanges. The interrelatedness of standardization and commodification, accordingly, could be observed in ancient times, when, for instance, Mediterranean cities appear to have adopted shared weight standards for coins to facilitate and expand trade (Psoma 2015). At the same time, however, it is undeniable that the interrelated projects of standardizing and commodifying "virtually every aspect of social life" were distinctively modern (Polanyi 1944; Busch 2011, p. 83). Some of the most powerful prostandardization rhetoric originated in the 18th and 19th centuries, the same period in which a new logic of commodification and fiction of the "self-regulating market" was invented and popularized and the industrial factory appeared as a "triumph of standardization" (Busch 2011, p. 120).

The slow but eventually widespread introduction of the metric system offers an important example illustrating how a zeal for standardization could accompany an advocacy for free-market policies. In revolutionary France, the first country to adopt metric measurement units, scientists partly embraced the system as an instrument for facilitating long-distance trade – a rational commercial language equivalent to the mathematical language popular in the physical sciences and natural philosophy of the time (Alder 1995, pp. 54–59). However, as historian Kenneth Alder (1995, p. 59) has shown, the aim of creating a well-integrated free market economy backfired, and the French government ended up reviving "the distinction, familiar to the old regime, between the regulated public *marketplace* (limited in time and place, so that all might have equal access) and the chaotic *free market*". Although opponents of the metric system strategically highlighted such setbacks and failures, the potential of metric units to enable commercial exchanges across large geographical distances contributed to the system's persistent appeal. These envisioned exchanges, for instance, included exports of agricultural commodities from and within post-independence Brazil, a relatively understudied example of a country whose scientists and engineers successfully advocated for metrification in the 19th century (Hanley 2022).

The history of scientific research technology, including that of laboratory animals, is another rich area for exploring in which circumstances and to what extent standardization has led to commodification (and vice versa). The scientific and medical value of "model organisms" such as the *Drosophila melanogaster* fruit fly and the Wistar rat partly results from the fact that these were "standard" animals that could serve as reference points in biological research and were relatively easy to reproduce and exchange (Clause 1993; Kohler 1994; Leonelli and Ankeny 2013). A laboratory animal that did not undergo genetic and/or pathogenic standardization would have been more difficult to commodify. However, even though it is currently possible for scientists worldwide to "choose a standard animal suitable for their purposes as easily as they might choose any other piece of technology or equipment from a trade catalogue" (Kirk 2010, p.92), standardization did not necessarily lead to commodification. As Robert Kohler (1994, Chapter 5) has argued, for instance, Thomas H. Morgan and his early-20th-century research group at Columbia University in New York initially treated their *Drosophila* stock as a common property and freely shared mutants with colleagues at other institutions, thus increasing the visibility and acceptance of their innovative research program and genetic mapping techniques. Although Morgan's group asked for a small financial compensation for supplying *Drosophila* for non-reciprocal teaching purposes, the researchers' moral rules and conventions would have made it unacceptable to run a for-profit fruit fly business (Kohler 1994, p.144). All the same, by the 1930s, commercial suppliers had entered the field and turned *Drosophila* into an unambiguous commodity – a status it has retained ever since.

A "moral economy" similar to that of Morgan's research school also guided the geneticist Clarence Cook Little of the Jackson Laboratory at Bar Harbor in Maine in promoting inbred mice as a research tool in the 1920s and early 1930s. Owing to financial pressures resulting from the Great Depression, however, Little ended up completing his first "mouse sales deal" in early 1933 (Rader 2004, p.144). Although Little himself did not realize this at the time, this half-a-dollar-a-piece sale of 500 mice to the Public Health Service started a chain of events that would profoundly affect the mission of his lab. As historian Karen Rader (2004, p. 134) has put it, the Jackson Lab primarily became a "mouse factory, not a research factory", and it set up a "more elaborate system of production and circulation aimed at monopolizing the scientific market for mice". The laboratory would indeed become a near-monopolist, dramatically scaling up its inbreeding work to the point where it was selling over 2 million genetically standardized "JAX" mice per year to outside biomedical researchers. As a result of its changed priorities, the lab had become known as "the bureau of mouse standards" by the 1950s (Rader 2004, p. 8). This perception nicely captured how commodification had fostered standardization in this case, yet it also hindered Little's efforts to attract funding for original research in mammalian genetics.

The entry of new types of public and private actors into the business of producing model organisms turned various additional animals into scientific commodities. In mid-20th-century Britain, for instance, the entrepreneurial director of the Laboratory Animals Bureau of the Medical Research Council (founded in

1947) managed to place his institution at the center of a newly constituted national market for generic lab animals by deploying a standardizing discourse. As historian Robert Kirk (2010) has argued, this discourse had substantial internationalizing power, as evident in the formation of an International Committee for Laboratory Animals in 1956 and the global emergence of laboratory animal science as a new scientific discipline (see also Druglitrø and Kirk (2014)). Government and academic scientists were not the only participants in this discipline; multinational biomedical enterprises also emerged as an important new actor involved in the study and inbreeding of standardized laboratory animals in the post-World War II decades. Among other things, companies such as Marshall Farms solidified the status of beagle dogs as standard animals for toxicological studies and, rather than relying on small commercial breeders or factory-like government or academic laboratories, initiated the in-house production of beagles on a large industrial scale (Bolman 2022).

The use of dogs as research animals, however, has been much more controversial than the use of rats and mice, and the beagle example also highlights the (potential) reversibility of commodification processes and the significance of standards in this respect. According to the philosopher of technology Paul B. Thompson (2012, p. 102), "the rise of technical standards for animal welfare can be seen as an attempt to discipline the unwanted consequences of market forces". Animal welfare – a concept popularized from about the 1960s (Woods 2012) – is an area where governments worldwide have set *de jure* standards rather than relied on voluntary consensus standards (e.g. MacArthur Clark and Sun (2020)). Journalists and activist groups such as the Beagle Freedom Project, however, have argued that these standards do not go far enough and that "all forms of animal exploitation" should end (Greenwald and Woodhouse 2018); see also Bolman (2022, p. 173). Importantly, these activists do not oppose the commodification of animals per se; to my knowledge, no one objects to selling beagles as pets. The standardization and commodification of dogs as laboratory animals, however, is bound to continue to arouse resistance, and the banning of great ape research in multiple countries demonstrates that decommodification is a real possibility here.

7.4. Conclusion

The study of technical standards and standardization processes is a subject that, owing to its economic, political, sociocultural, scientific and technological dimensions, almost necessitates a cross-disciplinary perspective. Historians of technology and economic historians have taken a range of different approaches, usually drawing at least to a limited extent on each other's work. Many historians working at the intersection of the history of technology and economic and business history have, in fact, addressed the significance of technical standards in their work, as this chapter has illustrated in a comprehensive but by no means exhaustive manner. This rich literature reveals that, despite the application of distinctive research methods and perspectives and explicit disagreements on how to make sense of artifacts such as the QWERTY/AZERTY keyboard, there has been enough common ground to allow for productive knowledge exchanges. Moreover, as I have argued in this chapter, technical standards should remain on historians' agenda in a time when many scholars have redefined themselves as historians of capitalism and be considered carefully in research on commodification and decommodification processes.

7.5. References

Alder, K. (1995). A revolution to measure: The political economy of the metric system in France. In *The Values of Precision*, Wise, M.N. (ed.). Princeton University Press, Princeton.

Barreyre, N. and Blin, A. (2017). À la redécouverte du capitalisme américain. *Revue d'histoire du XIXe siècle*, 54(1), 135–148.

Beckert, S. and Desan, C. (eds) (2018). *American Capitalism: New Histories*. Columbia University Press, New York.

Berg, M. (2007). The genesis of "useful knowledge". *History of Science*, 45(2), 123–133.

Bolman, B. (2022). Dogs for life: Beagles, drugs, and capital in the twentieth century. *Journal of the History of Biology*, 55(1), 147–179.

Brock, J.W. (1983). Persistent monopoly and the charade of antitrust: The durability of Kodak's market power. *University of Toledo Law Review*, 14(3), 653–684.

Busch, L. (2011). *Standards: Recipes for Reality*. MIT Press, Cambridge.

Busch, L. and Whyte, K.P. (2012). On the peculiarity of standards: A reply to Thompson. *Philosophy and Technology*, 25(2), 243–248.

Cahan, D. (1989). *An Institute for an Empire: The Physikalisch-Technische Reichsanstalt, 1871-1918*. Cambridge University Press, Cambridge.

Carlson, W.B. (1988). Academic entrepreneurship and engineering education: Dugald C. Jackson and the MIT-GE cooperative engineering course, 1907-1932. *Technology and Culture*, 29(3), 536–567.

Chandler, A.D., Jr. (1990). *Scale and Scope: The Dynamics of Industrial Capitalism*. Harvard University Press, Cambridge.

Clause, B.T. (1993). The Wistar rat as a right choice: Establishing mammalian standards and the ideal of a standardized mammal. *Journal of the History of Biology*, 26(2), 329–349.

Cochrane, R.C. (1966). *Measures for Progress: A History of the National Bureau of Standards*. National Bureau of Standards, Washington, D.C.

Cooper, C.C. (ed.) (1991). Special issue: Patents and invention. *Technology and Culture*, 32(4), 837–1101.

Cowan, R.S. (1983). *More Work for Mother: The Ironies of Household Technology from the Open Hearth to the Microwave*. Basic Books, New York.

Cronon, W. (1991). *Nature's Metropolis, Chicago and the Great West*. W.W. Norton & Company, New York.

de la Cruz-Fernández, P.A. (2021). *Gendered Capitalism: Sewing Machines and Multinational Business in Spain and Mexico, 1850-1940*. Routledge, New York.

David, P.A. (1985). Clio and the Economics of QWERTY. *The American Economic Review*, 75(2), 332–337.

David, P.A. (1997). Path dependence and the quest for historical economics: One more chorus of the ballad of QWERTY [Online]. Available at: https://ora.ox.ac.uk/objects/uuid:cfb6d53a-d0e3-48f9-bce3-b48ad3216c90 [Accessed 3 June 2023].

David, P.A. and Greenstein, S. (1990). The economics of compatibility standards: An introduction to recent research. *Economics of Innovation and New Technology*, 1(1/2), 3–41.

Deringer, W. (2018). Compound interest corrected: The imaginative mathematics of the financial future in early modern England. *Osiris*, 33(1), 109–129.

Druglitrø, T. and Kirk, R.G.W. (2014). Building transnational bodies: Norway and the international development of laboratory animal science, ca. 1956-1980. *Science in Context*, 27(2), 333–357.

Frohlich, X. (2022). Making food standard: The U.S. Food and Drug Administration's Food Standards of Identity, 1930s-1960s. *Business History Review*, 96(1), 145–176.

Gago, M. (2022). Moving coffee from the forests of colonial Angola to the breakfast tables of main street America, 1940-1961. In *Knowledge Flows in a Global Age*, Krige, J. (ed.). University of Chicago Press, Chicago.

Galambos, L. (1983). Technology, political economy, and professionalization: Central themes of the organizational synthesis. *Business History Review*, 57(4), 471–493.

Gardey, D. (1999). The standardization of a technical practice: Typing (1883-1930). *History and Technology*, 15(4), 313–343.

Gaudillière, J.P. and Löwy, I. (eds) (1998). *The Invisible Industrialist: Manufactures and the Production of Scientific Knowledge*. St. Martin's Press, New York.

Gooday, G. (2004). *The Morals of Measurement: Accuracy, Irony, and Trust in Late Victorian Electrical Practice*. Cambridge University Press, Cambridge.

Graham, M.B.W. (2008). Henry P. Gage: Entrepreneurial standards setter for corning glass works, 1911-1947. *Entreprises et Histoire*, 51, 28–43.

Greenwald, G. and Woodhouse, L.A. (2018). Bred to suffer: Inside the barbaric US industry of dog experimentation [Online]. Available at: https://theintercept.com/2018/05/17/inside-the-barbaric-u-s-industry-of-dog-experimentation [Accessed June 2023].

Hall, D. (2023). "Commodification of everything" arguments in the social sciences: Variants, specification, evaluation, critique. *Environment and Planning A: Economy and Space*, 55(3), 544–561.

Hanley, A.G. (2022). Men of science and standards: Introducing the metric system in nineteenth-century Brazil. *Business History Review*, 96(1), 17–45.

Hearty, R.J (2023). The pollution experts: Engineers, biologists, and the problem of water quality in rivers of the United States, 1935 to 1972. PhD Thesis, Johns Hopkins University, Baltimore.

Hilt, E. (2017). Economic history, historical analysis, and the "new history of capitalism". *The Journal of Economic History*, 77(2), 511–536.

Hounshell, D.A. (1984). *From the American System to Mass Production, 1800-1932: The Development of Manufacturing Technology in the United States*. JHU Press, Baltimore.

Hui, A. (2012). Sound objects and sound products: Standardizing a new culture of listening in the first half of the twentieth century. *Culture Unbound*, 4(4), 599–616.

Hunt, B.J. (1994). The ohm is where the art is: British telegraph engineers and the development of electrical standards. *Osiris*, 9, 48–63.

Inkster, I. (2006). Potentially global: "Useful and reliable knowledge" and material progress in Europe, 1474-1914. *The International History Review*, 28(2), 237–286.

Kadiyali, V. (1998). Eastman Kodak in the photographic film industry: Picture imperfect? In *Market Dominance*, Rosenbaum, D. (ed.). Greenwood Press, Westport.

Kirk, R.G.W. (2010). A brave new animal for a brave new world: The British Laboratory Animals Bureau and the constitution of international standards of laboratory animal production and use, circa 1947-1968. *Isis*, 101(1), 62–94.

Kocka, J. (2016). *Capitalism: A Short History*. Princeton University Press, Princeton.

Kocka, J. and van der Linden, M. (eds) (2016). *Capitalism: The Reemergence of a Historical Concept*. Bloomsbury Academic, London.

Kohler, R.E. (1994). *Lords of the Fly: Drosophila Genetics and the Experimental Life*. University of Chicago Press, Chicago.

Lamoreaux, N.R., Raff, D.M.G., Temin, P. (2003). Beyond markets and hierarchies: Toward a new synthesis of American business history. *The American Historical Review*, 108(2), 404–433.

Lauer, J. and Lipartito, K. (eds) (2021). *Surveillance Capitalism in America*. University of Pennsylvania Press, Philadelphia.

Lécuyer, C. (1998). Academic science and technology in the service of industry: MIT creates a "permeable" engineering school. *The American Economic Review*, 88(2), 28–33.

Lee, J., Veloso, F.M., Hounshell, D.A., Rubin, E.S. (2010). Forcing technological change: A case of automobile emissions control technology development in the US. *Technovation*, 30(4), 249–264.

Leonelli, S. and Ankeny, R.A. (2013). What makes a model organism? *Endeavour*, 37(4), 209–212.

Levy, J. (2017). Capital as process and the history of capitalism. *Business History Review*, 91(3), 483–510.

Lundgreen, P. (1997). Handling risk: Expertise and regulatory politics in Germany, 1870-1913. *IEEE Technology and Society Magazine*, 16(1), 16–22.

MacArthur Clark, J.A. and Sun, D. (2020). Guidelines for the ethical review of laboratory animal welfare People's Republic of China national standard GB/T 35892-2018. *Animal Models and Experimental Medicine*, 3(1), 103–113.

MacKenzie, D. (1984). Marx and the machine. *Technology and Culture*, 25(3), 473–502.

MacKenzie, D. (2015). Mechanizing the Merc: The Chicago mercantile exchange and the rise of high-frequency trading. *Technology and Culture*, 56(3), 646–675.

McCraw, T.K. (2007). *Prophet of Innovation: Joseph Schumpeter and Creative Destruction*. Belknap Press of Harvard University Press, Cambridge.

Meister, A.-M. (2020). Ernst Neufert's "Lebensgestaltungslehre": Formatting life beyond the built. *BJHS Themes*, 5, 167–185.

Mihm, S. (2022). Inching toward modernity: Industrial standards and the fate of the metric system in the United States. *Business History Review*, 96(1), 47–76.

Moguen-Toursel, M. (2008). Éditorial. Standardisation : l'apport des approches historiques. *Entreprises et histoire*, 51(2), 6–9.

Mokyr, J. (2002). *The Gifts of Athena: Historical Origins of the Knowledge Economy*. Princeton University Press, Princeton.

Moseley, R. (1978). The origins and early years of the National Physical Laboratory: A chapter in the pre-history of British science policy. *Minerva*, 16(2), 222–250.

Murphy, C. and Yates, J. (2009). *The International Organization for Standardization (ISO): Global Governance through Voluntary Consensus*. Routledge, London.

Noble, D.F. (1977). *America by Design: Science, Technology, and the Rise of Corporate Capitalism*. Knopf, New York.

Ordóñez, J. and Sánchez, A. (2020). Introduction: Standards in the history of contemporary science. *HoST-Journal of History of Science and Technology*, 14(1), 1–5.

Polanyi, K. (2001). *The Great Transformation: The Political and Economic Origins of Our Time*. Beacon Press, Boston.

Psoma, S.E. (2015). Choosing and changing monetary standards in the Greek world during the archaic and the classical periods. In *The Ancient Greek Economy*, Harris, E.M., Lewis, D.M., Woolmer, M. (eds). Cambridge University Press, New York.

Puffert, D.J. (2009). *Tracks across Continents, Paths through History: The Economic Dynamics of Standardization in Railway Gauge*. University of Chicago Press, Chicago.

Rader, K.A. (2004). *Making Mice: Standardizing Animals for American Biomedical Research, 1900-1955*. Princeton University Press, Princeton.

Reinstaller, A. and Hölzl, W. (2009). Big causes and small events: QWERTY and the mechanization of office work. *Industrial and Corporate Change*, 18(5), 999–1031.

Rieppel, L., Lean, E., Deringer, D. (2018). Introduction: The entangled histories of science and capitalism. *Osiris*, 33(1), 1–24.

Rood, D. (2017). *The Reinvention of Atlantic Slavery: Technology, Labor, Race, and Capitalism in the Greater Caribbean*. Oxford University Press, New York.

Rosenberg, N. (1982). *Inside the Black Box: Technology and Economics*. Cambridge University Press, Cambridge.

Rosenthal, C. (2018). *Accounting for Slavery: Masters and Management*. Harvard University Press, Cambridge.

Russell, A.L. (2005). Standardization in history: A review essay with an eye to the future. In *The Standards Edge*, Bolin, S. (ed.). University of Michigan Press, Ann Arbor.

Russell, A.L. (2008). Dot-org entrepreneurship: Weaving a web of trust. *Entreprises et histoire*, 51(2), 44–56.

Russell, A.L. (2014). *Open Standards and the Digital Age: History, Ideology, and Networks*. Cambridge University Press, New York.

Russell, E. (2022). Capitalism matters: How financial and technological innovations shaped U.S. telegraphs, 1845-1860. *Technology and Culture*, 63(1), 31–60.

Russell, A.L. and Vinsel, L. (2018). After innovation, turn to maintenance. *Technology and Culture*, 59(1), 1–25.

Schäfer, D. and Valeriani, S. (2021). Technology is global: The useful and reliable knowledge debate. *Technology and Culture*, 62(2), 327–347.

Schaffer, S. (1992). Late Victorian metrology and its instrumentation: A manufactory of Ohms. In *Invisible Connections: Instruments, Institutions, and Science*, Bud, R. and Cozzens, S. (eds). SPIE Optical Engineering Press, Bellingham.

Schakenbach Regele, L. (2023). A brief history of the history of capitalism, and a new American variety. doi: 10.1017/eso.2023.35.

Schwab, K. (2017). *The Fourth Industrial Revolution*. Crown Business, New York.

Scranton, P. (1997). *Endless Novelty: Specialty Production and American Industrialization, 1865-1925*. Princeton University Press, Princeton.

Scranton, P. and Fridenson, P. (2013). *Reimagining Business History*. JHU Press, Baltimore.

Shapiro, C. and Varian, H.R. (1999). The art of standards wars. *California Management Review*, 41(2), 8–32.

Sinclair, B. (1969). At the turn of a screw: William Sellers, the Franklin Institute, and a standard American thread. *Technology and Culture*, 10(1), 20–34.

Singerman, D. (2018). Sugar machines and the fragile infrastructure of commodities in the nineteenth century. *Osiris*, 33(1), 63–84.

Slaton, A. (2001). "As near as practicable": Precision, ambiguity, and the social features of industrial quality control. *Technology and Culture*, 42(1), 51–80.

Slaton, A. and Abbate, J. (2001). The hidden lives of standards: Technical prescriptions and the transformation of work in America. In *Technologies of Power: Essays in Honor of Thomas Parke Hughes and Agatha Chipley Hughes*, Allen, M.T. and Hecht, G. (eds). MIT Press, Cambridge.

Spencer, C. and Temple, P. (2016). Standards, learning, and growth in Britain, 1901-2009. *The Economic History Review*, 69(2), 627–652.

Spiekermann, U. (2018). *Künstliche Kost Ernährung in Deutschland, 1840 bis heute*. Vandenhoeck & Ruprecht, Göttingen.

Spiekermann, U. (2011). Redefining food: The standardization of products and production in Europe and the United States, 1880-1914. *History and Technology*, 27(1), 11–36.

Stango, V. (2004). The economics of standards wars [Online]. Available at: https://doi.org/10.2202/1446-9022.1040 [Accessed 19 June 2003].

Sumner, J. and Gooday, G.J.N. (2008). Introduction: Does standardization make things standard? *History of Technology*, 28, 1–13.

Thompson, G.V. (1954). Intercompany technical standardization in the early American automobile industry. *The Journal of Economic History*, 14(1), 1–20.

Thompson, P.B. (2012). "There's an app for that": Technical standards and commodification by technological means. *Philosophy and Technology*, 25(1), 87–103.

Timmermans, S. and Almeling, R. (2009). Objectification, standardization, and commodification in health care: A conceptual readjustment. *Social Science and Medicine*, 69(1), 21–27.

Timmermans, S. and Epstein, S. (2010). A world of standards but not a standard world: Toward a sociology of standards and standardization. *Annual Review of Sociology*, 36(1), 69–89.

Usselman, S.W. (2002). *Regulating Railroad Innovation: Business, Technology, and Politics in America, 1840-1920*. Cambridge University Press, Cambridge.

Vermij, R. (2010). Scale increase and its dynamic. In *Technology and the Making of the Netherlands: The Age of Contested Modernization, 1890-1970*, Schot, J., Rip, A., Lintsen, H. (eds). Walburg Pers, MIT Press, Zutphen.

Vinsel, L. (2016). Virtue via association: The National Bureau of Standards, automobiles, and political economy, 1919-1940. *Enterprise and Society*, 17(4), 809–838.

Vinsel, L. (2019). *Moving Violations: Automobiles, Experts, and Regulations in the United States*. JHU Press, Baltimore.

Ward, J. (2019). Financing the information age: London TeleCity, the legacy of IT-82, and the selling of British telecom. *Twentieth Century British History*, 30(3), 424–446.

Woods, A. (2012). From cruelty to welfare: The emergence of farm animal welfare in Britain, 1964-1971. *Endeavour*, 36(1), 14–22.

Woods, B. and Watson, N. (2004). In pursuit of standardization: The British Ministry of Health's model 8F wheelchair, 1948-1962. *Technology and Culture*, 45(3), 540–568.

Yates, J. and Murphy, C.N. (2021). *Engineering Rules: Global Standard Setting since 1880*. JHU Press, Baltimore.

Yates, J. and Murphy, C.N. (2022). Introduction: Standards and the global economy. *Business History Review*, 96(1), 3–15.

Ye, S., Xiao, H., Zhou, L. (2018). Commodification and perceived authenticity in commercial homes. *Annals of Tourism Research*, 71, 39–53.

Zuboff, S. (2019). *The Age of Surveillance Capitalism: The Fight for a Human Future at the New Frontier of Power*. Public Affairs, New York.

8

History of Techniques and the Environment

Anaël MARREC
CHC, CNRS, Université Paris 1 Panthéon-Sorbonne, Paris, France

8.1. Introduction

> Against a backdrop of destruction of life and social structures by a technocratic elite concentrating power and wealth, we engineers understand that through our activities, we are contributing to the aggravation of ongoing environmental and social disasters [...] We are now deserting our professions, our studies, our positions, so that we can organize ourselves outside the systems that have shaped us, and the closed and hermetic world of engineers. Some knowledge and skills will have to be abandoned. Others will enable us to understand what we want to tackle, or can be redirected and put to use to support struggles.[1]

The call to "desert" the engineering profession from the prestigious agronomy school AgroParisTech, in Spring 2022, was one of a number of speeches made by students and young graduates of French engineering schools that received particular

---

1 Manifesto of the Désert'heureuses, a group of young engineering graduates. Available at: www.desertheureuses.noblogs.org/manifeste/ [Accessed 10 August 2023].

media coverage[2]. It expresses an uneasiness in the vocation of engineers, as well as a virulent critique of the role of techniques in the contemporary ecological crisis. These questions about industrial production methods are not isolated and are based on a damning observation: global warming and its many manifestations (droughts, heatwaves, fires, rising sea levels, etc.), the drastic decline in marine and terrestrial biodiversity, destruction of ecosystems through the extraction of raw materials and energy, soil erosion, various forms of pollution, epidemics, etc., are the result of increased environmental pressure from human societies (Bonneuil and Fressoz 2013), in which technical innovations play their part. Furthermore, these environmental problems are experienced with varying degrees of intensity across the planet's landscapes, and the most vulnerable populations are disproportionately exposed. Questions about techniques also focus on their potential as drivers for more environmentally friendly lifestyles and fairer societies. Proposals for societies based on "low tech" that promote autonomy and respect for the environment, as opposed to decontextualized, undemocratic and destructive "high tech", provide a good illustration of this (Abrassart et al. 2020). The latter highlights the fact that techniques are not neutral and that, depending on their form and purpose, they are embedded in social structures and environments in different ways. There is a long history of these questions in the contemporary era, which is expressed in the criticism of techniques and in the search for alternatives (Jarrige 2014; Lopez 2014). Thus, "it would obviously be absurd to set man against technique", and criticism has focused less on "the" technique itself than on its various forms and their social and environmental implications (Jarrige 2014, pp. 12–18).

The doubts and hopes raised by technology in the "Anthropocene" era, as well as the profound and numerous temporalities of technical and environmental transformations, call for a historical understanding of the relationships between technical change and ecological transformations (Demuth 2019; Van der Straeten and Weber 2025). This objective calls into question both the history of techniques, which seeks to understand techniques in society through history (Pestre and Cohen 1998), and environmental history, which aims to understand the dynamic links between societies and their environments in their material, institutional and ideal dimensions (Massard-Guilbaud 2007). The technique–environment link appears to be essential in the most common understanding of "techniques": whether considered through a nature/culture split or not, techniques have generally been defined as

2 For the call of AgroParisTech, see *Le Monde*, "Des étudiants d'AgroParisTech appellent à 'déserter' des emplois 'destructeurs'", published May 11, 2022 [Online]. Available at: www.lemonde.fr/planete/article/2022/05/11/des-etudiants-d-agroparistech-appellent-a-deserter-des-emplois-destructeurs_6125644_3244.html [Accessed 10 August 2023].

mediating elements between societies and their environments[3] or between "humans" and "non-humans"[4]. Techniques, however, go beyond the purely productive or utilitarian question, as they also translate "goals, values and regimes of power into practices and artifacts" (Bray and Hilaire-Pérez 2016). Thus, since the social turn of the 1980s, the history of techniques has tended to make environments a secondary object of study (Pritchard 2014). On the other hand, since the 1990s, environmental history has moved away from declinist narratives in which human action was necessarily destructive of a "nature" that was external to it, preferring to study environments not as an object but as "a field of force that passes through and is passed through by human processes and actions, and which appears through configurations whose history can be traced" (Parrinello 2018b). The aim of this historiographical field is not so much to examine a new object (the environment) but rather to bring about an "environmental eye" to ask "questions that traditionally belonged to economic history, social history, the history of public policy, the history of development or infrastructure" (Massard-Guilbaud 2007, p. 16). While "envirotechnical" approaches have been developed in the United States since the early 2000s (Pritchard 2014), few French historians have yet to explicitly take an environmental approach to the history of techniques.

This does not mean that the environment has been absent from the history of techniques, nor vice versa. The history of techniques has taken a keen interest in development and resource issues, without necessarily claiming to be environmental (Fournier 2003; Hilaire-Pérez and Verna 2016). Environmental history, however, which emerged in the United States in the 1970s and was established in France at the turn of the 2000s, was driven by researchers, including urban historians and historians of science and technology, with the former traditionally interested in territories and infrastructures and the latter in ways of thinking about and acting on nature (Fressoz et al. 2014). This chapter provides an overview of how the history of techniques and environmental history have dealt with the relationship between techniques and environments. To restrict the analysis and extend Sarah Pritchard's (2014) work concerning essentially American literature, it will look at French

---

3 In anthropologist Marcel Mauss's classic definition of a "traditional efficient act" (Mauss 1934, p. 9), the environment is not mentioned, but it is implicit in the idea of action and efficiency, which is largely based on mechanistic thinking: action has an exogenous element as an objective, to which it applies an impulse with a view to a project.

4 According to the philosopher and technical sociologist Bruno Latour, "technique is defined by the mediation of relationships between people, on the one hand, and between people, things and animals, on the other hand" (Latour 1993, p. 15).

literature while being attentive to international influences. After outlining how the history of techniques has considered environments and how environmental history has considered technical issues, this chapter will focus on the explicit junctions between these two approaches.

8.2. Environment in the history of techniques (1960s–1990s): natural background, social context, resources

In a recent interview, André Guillerme asserted that "there can be no isolated history of techniques: it is also environmental history" – to do otherwise "would be to sin by omission" (Fernandez and Guillerme 2018, p. 4). The French historian of techniques went back over his career and recalled his interest, as early as the 1980s[5], in the relationship between infrastructures and environments in cities but also his isolation on this subject until the late 1990s. While environmental history had been developing since the 1970s in the United States and the 1990s in Western Europe, it was slow to penetrate the French historiographical field. In 2007, Geneviève Massard-Guilbaud was still urging her colleagues in all fields of history to examine their objects with an "environmental eye", underlining the urgency of taking up these issues and breaking out of a form of "blindness" that was becoming unacceptable. Indeed, the impacts, power relations and perceptions of different forms of development and modes of resource mobilization and production in territories and on bodies, and across social groups, had a history that remained largely unwritten (Massard-Guilbaud 2007, p. 16). Indeed, in France, from the 1970s onward, "the social critique of productivism struggled [...] to articulate itself with the ecological question", as it was suspected of reactivating reactionary visions in the eyes of the human and social sciences (Jarrige 2017, p. 77). Historians therefore still measured "prosperity by the height of the chimneys and the blackness of the landscape" (Massard-Guilbaud 2002, p. 67). The history of techniques was also largely concerned with this observation. Despite the writing of great universal sagas of the history of technology from the 1960s onward, there was no work comparable to Lewis Mumford's *Technics and Civilization* (2016), which tackled the relationships between technology, social structures and environmental impacts. A number of French studies nevertheless stand out for their early interest in the relationship between societies and environments.

5 This was reflected in his first book, published in 1983, which examined the relationship between cities and water in Europe from antiquity to the 19th century (Guillerme 1983).

8.2.1. *History of techniques through great universal sagas: the environment as a natural backward*

In the 1970s, the history of techniques was still a relatively new field in France, one that was in the process of being structured, particularly with the establishment in 1962 of the Centre de documentation d'histoire des techniques (history of techniques documentation center) at the Conservatoire national des métiers, CNAM (National Conservatory of Trades), headed by Maurice Daumas (Larroque 1991; Guillerme 2007). In terms of scientific production, this period was marked by two encyclopedic works covering the history of techniques in different civilizations since Antiquity: those of Maurice Daumas (1962–1979) and Bertrand Gille (1978). Daumas set out to achieve the goal of a "technical history of techniques", formulated by Lucien Febvre in an issue of *Annales d'histoire économique et sociale* in 1935[6]. His work was thus deliberately internalist, describing technique as having its own logic. Gille similarly analyzed techniques within "technical systems", which he nevertheless described as constantly interacting with economic and social systems. Both of these major works have been recognized for their contributions. In particular, their examination over the long term revealed the specific temporalities of techniques and challenged the common idea of industrial revolutions (Fournier 2005; Hilaire-Pérez and Verna 2016). They encountered criticism, however, in particular for their propensity to extract technical phenomena from the social and economic spheres (Salomon 1980) and, later, for their ethnocentric and teleological presuppositions in a long line of evolutionary anthropology (Gras 1998; Carnino et al. 2023). They integrate environments in two main ways: as a constraint to be mastered and as a resource to be exploited through technological progress. In the Braudelian tradition, the environment is a natural backward, in other words, a framework with long-lasting temporalities that are external to those of societies (Massard-Guilbaud 2002). When the word "environment" is used, it refers to a social and technological context: the environment of a work organization (Daumas and Mioche 2004, p. 75). The technique–environment link described is one of conquest: technique enables societies to extricate themselves from natural hazards. There is no reflexivity about the underpinning representations of nature, while issues of environmental impacts and conflicts are not addressed.

---

6 Febvre called for the creation of a history of techniques, with the technical history of techniques as the first objective, the other two being a history of techniques linked to the history of science, and a history of techniques embedded in economic and social history (Febvre 1935).

8.2.2. *History of companies and social studies: social and economic context, instrumental materiality*

In the 1980s, the history focused on technical change and its actors, which were still mindful of the materiality of techniques but made environments a secondary issue in favor of the social construction of techniques. In 1980, the historian of science Jean-Jacques Salomon, commenting on the works of Daumas and Gille, provided a good illustration of the reduction of the technical environment to social and economic contexts, that of designers and engineers, and of the marginalization, between the 1980s and 1990s, of the history of work and the working class, which was correlated with deindustrialization (Hatzfeld et al. 2016). In his view, "French studies of the industrial revolution have focused on work rather than technique, on the worker who suffers rather than the one who creates, on the craftsman but not on the engineer, in such a way that the history of techniques was initially covered and, more often than not, obscured by the workers' movement and social change" (Salomon 1980, pp. 455–456). He therefore calls for a shift away from material questions toward a conceptual and intellectual history of techniques, which is largely what French research into the history of techniques would do over the next two decades.

In the 1980s and 1990s, French business historians took a renewed interest in innovation and the history of large technological systems such as railways and electricity, particularly through the work of François Caron and the *Comité d'histoire de l'électricité* (Griset and Bouvier 2012). Influenced by Schumpeterian thought, this history considers the phenomena of technical destruction and creation guided by the market and the creativity of entrepreneurs. It considers the social space of techniques, constituted as systems, owing to the triptych state/company/consumer. While attentive to the conditions of deployment of technological systems, this history considers the environment as a socioeconomic environment constituted by companies, engineers, technicians and consumers. As the business historians Philippe Mioche and Jean-Claude Daumas noted in 2004, calling for a renewal of their field (Daumas and Mioche, p. 69), the environment has no place in it, except in the form of technical obstacles (to be circumvented) or social "resistance" (to the dangers of electricity or to technical change), which will sooner or later have to give way to the necessary process of modernization. Very active in Anglo-American literature, social construction of technology (SCOT) took shape in northern Europe and the United States[7] and was influenced by science and technology studies (Pestre and Cohen 1998). The main effect of the SCOT approach is to "move the history of

7 See, in particular, Mackenzie and Wajcman (1999) and Bijker et al. (1987).

techniques towards a closer focus on actors and the resources[8] at their disposal" (Picon 1995). It contributes to challenging progressive and deterministic accounts of the history of techniques, blurring the boundaries between social actors and artifacts, and criticizing the notion of efficiency (Picon 1995): technical systems and societies form "seamless webs" (Hughes 1989, pp. 76–77). The contribution of social studies to the history of techniques in terms of social analysis has been less spectacular than in the history of science, which was more inclined to extract its objects of study from economic and social contexts (Picon 1995; Pestre and Cohen 1998). As Sarah Pritchard noted, SCOTs also tended to exclude environments from their analyses (Pritchard 2014). Indeed, the materiality of techniques and environments mattered less than the "concrete forms of human and social shaping of the object" (Pestre and Cohen 1998, pp. 729–730). The way historians of science Yves Cohen and Dominique Pestre described the realities (the "terrain") with which techniques are faced is characteristic of the absence of environments in social studies: "The *terrain* is competition, rivalry, war, at all levels, between individuals, between companies, between groups, between countries" (Pestre and Cohen 1998, pp. 729–730). However, by including natural elements in the human/non-human equation, Bruno Latour and Michel Callon's actor–network theory foreshadowed a wealth of developments in environmental history in the 2000s with "hybrid" approaches.

8.2.3. *History of engineers, cities and industry: pollution, waste, resources*

In the 1980s–1990s, two types of approaches gave a prominent place to environments. The first focused on engineers in France in the Middle Ages. This field makes it possible to pursue the study of techniques as a field of knowledge with its own historicity, which cannot be reduced to the application of science (Hilaire-Pérez 2019). Although these works do not claim to do so, they also make valuable contributions to the history of society–environment relations from a cultural and political perspective through the study of cognitive regimes and values at work in the transformation of territory by a social group that was of major importance in the construction of governments. Anne Blanchard (1979), Hélène Vérin (1993) and Antoine Picon (1988, 1992) describe the constitution and changes of this profession, its rationality and its social role (Fournier 2005). Picon in particular highlighted the task of the 18th century State engineers to measure and understand the national landscape in order to homogenize it and optimize the control of resources, goods and people (Picon 1988, 1992).

8 Not in the sense of a "natural resource", but of intellectual, material and social resources.

Finally, before the "environmental turn" of the 2000s in France, some works in urban, industrial, development and energy history stand out for the explicit place they gave to environmental issues, addressing the question of pollution and urban waste, resource governance and the relationship between modes of energy mobilization and environments. In the field of urban history, André Guillerme has worked on water infrastructures from the Middle Ages to the present day and on contemporary underground infrastructures since the 1980s (Guillerme 1983; Barles and Guillerme 1995). His work at the CNAM's *Centre d'histoire des techniques* (Center for the History of Techniques)[9], which was renamed the *Centre d'histoire des techniques et de l'environnement* (Center for the History of Techniques and the Environment) in 2002 under his direction from 1996 to 2012 (Guillerme 2007), was foundational in the development of French environmental history (Massard–Guilbaud 2002, p. 70).

In the field of resource management, collective work devoted to forests and their management in the modern and contemporary periods was written in 1990 (Woronoff 1990) and provided crucial information on technical issues (Hilaire-Pérez and Verna 2016). It discusses industrial responses to state regulations, such as improving forge performance and silvicultural management techniques, as well as protests against industries by rural and urban wood users, who feared the depletion of resources.

In the history of energy, a trio of Marxist-inspired historians of science and economics delivered a study on the ways in which energy has been used in different civilizations since Antiquity (Debeir et al. 2013). Techniques are materialized by "converters" organized into "energy systems" that determine a certain use of resources and a certain type of production (Debeir et al. 2013, p. 25). This work is remarkable for its attempt to grasp the material conditions of societies, their environmental footprint and the power relations induced by energy systems, leading them to later become an inspiring source for the environmental history of energy (Massard-Guilbaud 2018; Bécot et al. 2023). It is also characterized by a certain form of "declinism" in its narrative (one also present in American environmental history at the same time), focused on civilizations and the inescapable predation of their energy systems until resources are exhausted. A detour into the American environmental history of the 1970s–1980s and the role of techniques in its narratives will enable us to look back at a turning point in environmental history in the United States in the 1990s, which preceded the establishment of environmental history in France.

9 The new name for the *Centre de documentation d'histoire des techniques* (History of techniques documentation center).

8.3. Environmental history and the history of techniques

In environmental history, techniques have occupied a certain place as key elements of society/environment relations, both as an agent of transformations and their measurement, resource governance and the management of the undesirable consequences of human activities (Fressoz et al. 2014). In France, however, they have rarely been a specific subject of environmental history. A number of avenues can be explored to answer a question that was put forward at the first *Réseau universitaire de chercheurs et chercheuses en histoire environnementale*, RUCHE (University Network of Environmental Researchers) symposium in 2012: "What place has the history of techniques taken, and may still take in the future, in the field of environmental history?"[10], and how can the environmental perspective renew the history of techniques? Pritchard proposed a synthesis of essentially American works following the "envirotechnical" approach in the wake of hybrid approaches. Their guiding principle is to consider that "the 'biological' organisms, 'natural' environments, and 'technological' objects are in fact envirotechnical entities: hybrid materials of what is commonly referred to as nature and technology" (Pritchard 2014, p. 244). The goal here is to recall how the "first two generations" of environmental history integrated techniques and the emergence of envirotechnical approaches in the United States, as well as the different directions taken in France by what might be qualified as the environmental history of techniques.

8.3.1. *Techniques in American environmental history: from the destruction of nature to hybrids*

Environmental history has gone through two historiographical "generations" in the United States, the first being in the 1970s–1980s and the second in the 1990s (Fressoz et al. 2014; Parrinello 2018b). The disruption between the first two generations involved opposing conceptions of the environment and the place of human societies, which also induced radically different ways of considering techniques.

The "first generation" emerged in the United States in the wake of ecology and nature conservation movements. Inheriting a typically North American vision of "nature", it focused on environments that would be free of any human presence, recounting their decline through contact with colonizing societies and excluding

10 RUCHE, "Techniques et environnement", Conference announcement. Published Monday, July 9, 2012 [Online]. Available at: www.calenda.org/209206 [Accessed 11 August 2023].

anthropized environments. Techniques were thus reduced to destructive agents (Pritchard 2014). This history was strongly criticized by historians, who founded the "second generation" of environmental history. Urban historians such as William Cronon, Joel A. Tarr and Martin V. Melosi were particularly active in claiming the role of cities in environmental history (Cronon 1991; Melosi 1994; Journal of Urban History 1994). They emphasized the politically and socially situated vision of nature that resulted in a narrow conception of environments, typical of white middle classes concerned with preserving a mythified wilderness (Massard-Guilbaud 2007, p. 9). In addition to lacking reflexivity, this posture ignored the organic links that bind cities and the countryside and neglected the environment of a major part of the world's population, as well as urbanization, a crucial phenomenon in contemporary environmental transformations.

The second generation redefined the environment of environmental history as a hybrid object whose anthropized nature is no longer prohibitive; quite the contrary (Parrinello 2018b). Influenced by hybrid thinkers such as Latour or Harraway, Cronon, Tarr and Melosi, who were primarily urban historians of techniques and infrastructures, consider environments as singular configurations of the social, the technical and the "non-human"[11].

In the wake of the cultural turn, historians of this generation were also more interested in the representations of the environments underpinning economic activities. Techniques thus endorsed a singular role. To "overcome the opposition between progress and decline, [they took] three forms (sometimes combined): man shapes nature; the unexpected consequences of his actions push him into a cycle of technical transformations; the inextricably linked environmental and social changes create losers and winners" (Fressoz et al. 2014, pp. 11–20). Work such as that of Richard White on the Columbia River is characteristic of this shift from the wilderness to a hybrid environment by regarding the river as an "organic machine" (White 1995).

8.3.2. *An environmental history of technology: "envirotechnical" approaches*

The turn toward hybridity in environmental history led some historians of techniques to explicitly raise the question of the relationship between environmental

11 This turning point in environmental history also reflects the concurrent renewal of environmental ethics. The latter now promotes approaches that are less centered on the idea of an "intrinsic value" of nature, but are more pragmatic and contextual (Larrère 2010), "recognizing only local ecological 'objects', which are always specific and intrinsically transitory" (Blandin 1992).

history and the history of techniques in the early 2000s (Stine and Tarr 1998). Hughes himself, who had been a driving force behind the social turn in the history of techniques, proposed reviewing the history of techniques as the history of "ecotechnological environments" (Hughes 2004, p. 14). The term "envirotech" emerged in the 2000s around a group of historians at the Society for the History of Technology (SHOT), who claimed the junction between the history of techniques and environmental history (Russ and Cutcliffe 2010; Russell et al. 2011). For Pritchard, who set out to define a terminology for envirotechnical analysis, the approach essentially aims to include non-human components (biological and ecosystemic) in sociotechnical analyses such as Hughes's Large Technological Systems (LTS) or Hecht's technopolitics (Hecht 2014; Pritchard 2014). The notion of "envirotechnical systems" thus aims to "deepen the concept of the technological system, which was originally conceptualized as purely socio-technical, so as to include the environment" (Pritchard 2014, p. 245). The strength of this proposal lies in its ability to grasp the different configurations that can exist between environments and techniques while avoiding the pitfall of a deterministic history of techniques or a "declinist" history of environments. It considers the historicity of technique–environment links, particularly the complexity of their temporalities, as well as the inscription of technologies in landscapes as inhabited spaces, places of technological projections and infrastructures, and ecosystems. Hybrid approaches can run the risk of depoliticization, as with actor–network theory, criticized for dissolving social, environmental and technical factors into a chain of undifferentiated actors, making power relationships invisible (Parrinello 2018b). Pritchard, for example, defends the idea of a fundamental ambiguity in all techniques, which would have the potential to destroy as well as protect and repair environments. Some examples can be debated, such as the atom, which would have enabled better knowledge of environments, or the automobile, which would have enabled the middle classes to know nature and protect it (Pritchard 2014, pp. 235–239), if we consider the environmental and social trap in which these two technologies currently place societies. However, attention to power as simultaneously determined by the laws of nature, technical systems and interrelated social structures (Russell et al. 2011), as well as the notion of an "envirotechnical regime" (Pritchard 2011), makes it possible to position environments as "battlefields where multiple power relations are produced and transformed" (Parrinello 2018b).

Envirotechnical approaches have thus fueled a number of works, mainly in the United States, but also internationally, moving investigations into cities and their infrastructures, rivers and their development (Pritchard 2004, 2011; Parrinello 2018a), forests (Gardner 2009), mines (LeCain 2009), space and its waste (Rand 2017), coastlines (Booker 2013; Parrinello et al. 2022), and objects such as animal bodies, artificial respirators, the recycling industry (LeCain 2005), and energy converters (Russell et al. 2011; Marrec 2022). Two observations can be made about envirotechnical approaches: on the one hand, they have tended to focus on modern periods, and, on the other (which is related), they consider "technologies" not "techniques", ignoring techniques that do not fall under the application of science (Carnino et al. 2023). Furthermore, the attention to landscapes and their human/non-inhabitants, which greatly enriches the understanding of technical phenomena as close as possible to their stakeholders, runs the risk of confining the field to local areas. Current research dynamics, both in environmental history and in the history of techniques within the European and French frameworks, have opened new prospects for the long term, as well as for a multiplicity of spatial and temporal scales, all the way up to global issues.

8.3.3. *An environmental history of techniques?*

French research has rarely adopted explicit crossovers leading to specific terminology, such as those found in the envirotechnical literature. Environmental history began to emerge at the turn of the 2000s in the wake of the "second generation". Techniques are rarely specific objects of analysis, although they often take center stage. In addition to the history of techniques and urban environments[12], environmental history was built by dealing with productive and industrial activities, infrastructures and technical innovations over the long term, industrial pollution, wetland drying or the management of forest resources (Bernhardt and Massard-Guilbaud 2002; Massard-Guilbaud 2002, pp. 68–71; Fressoz et al. 2014). Studies on industrial pollution have been deepened and extended by important works studying their regulations as well as those of risk, describing "disinhibition devices" most often to benefit industrialists to the detriment of vulnerable populations (Massard-Guilbaud 2010; Leroux 2011; Fressoz 2012; Jas and Boudia 2013; Jarrige and Leroux 2017), as well as developments and their social and environmental consequences (Massard-Guilbaud and Fournier 2016), or occupational health (Rainhorn 2019; Bécot 2022a). The long-standing interest in environments shown by historians of techniques, particularly in the ancient, medieval and early modern periods, probably explains the difficulty, as in other historical fields, of seeing the environmental "eye" as something new (Massard-Guilbaud 2002). This reluctance also probably stems from concern of not reducing techniques to productive or industrial activities and the material counterpart of technoscience, in contrast to an environmental history largely represented, particularly in North America, by modern periods.

12 Around the work of André Guillerme at the CNAM, as well as Sabine Barles, on urban waste management (Barles 2005).

Nevertheless, the desire to cross the two historiographical fields has been evident for some fifteen years. In 2005, Patrick Fournier stressed that the history of techniques was the "key" to the history of relations between societies and their environments and provided several examples of work on the history of hydraulic techniques in the modern era (Fournier 2005). François Jarrige's work on the history of "technocritics" in the contemporary era was particularly innovative, studying both social contestations of technical modernization in Europe and their motivations, including environmental ones, and showing the contingency of "technical trajectories" (Jarrige 2009, 2014). In September 2012, during the young RUCHE's abovementioned first conference, French and European historians of techniques and the environment explicitly addressed the issue, focusing on hydraulic engineering and its environmental impact, industrial processes and their pollution, fishing practices in the face of the depopulation of fish-farming rivers, and the marginalization of ancient agricultural practices[13].

The renewed interest in environmental issues in industrial and business history has also benefited the history of techniques, in which economic historians have played an important role for a long time (Hilaire-Pérez and Verna 2016). Interest in themes such as companies and global warming (Adderi and Kartel 2017) or the circular economy (Marty and Druelle-Korn 2023), as well as industrial territories and their environments (Bartolotti 2020; Chatterji 2022), shed light on the power relationships at play in technological trajectories and their environmental counterparts, whether they are invisible, tolerated or instrumentalized. They also show the changing relationships between social groups and the infrastructures that surround them, as well as the economic opportunities that arise from industrial overflow through innovations such as the reuse of industrial residues. From this point of view, the literature on green business, which emerged in the United States around historian Geoffrey Jones, offers inspiring perspectives by questioning the relationships between profits and the environment around "green" entrepreneurs and innovations (Jones 2019).

Current research in the history of techniques and environmental history still offers fruitful prospects for crossovers, with spatial and temporal complexification linked to the global nature of contemporary environmental and technical issues. The history of techniques is undergoing a renewed dynamic through global approaches,

13 RUCHE, "Techniques et environnement", Conference organized on September 6 and 7, 2012 in Paris. Available at: www.leruche.hypotheses.org/podcasts/le-colloque-techniques-et-environnement-en-podcast [Accessed 11 August 2023].

as evidenced by French and European publications, events and research projects[14]. This work opens up the history of techniques to territories and processes that were previously ignored due to methodological presuppositions stemming from persistent ethnocentrism (Bray and Hilaire-Pérez 2016). A number of historical phenomena linked to techniques were reduced: modernization, imposition from Europe of certain modes of production; techniques, "technology" and the applications of technoscience; and territories to national frameworks, leading to the identification of technological "gaps". The global history of techniques proposes a number of shifts: the study of two-way circulations, instead of "technical transfers" from Europe. This opens up to areas outside Europe and North America and makes it possible to examine them in non-national frameworks[15], owing not only to the study of circulations between territories but also to regional approaches or notions such as the "technological landscape" (Van der Straeten 2019). Editorial projects on the global history of techniques over the long term (Bray and Hilaire-Pérez 2019; Carnino et al. 2023) have particularly emphasized the need to focus on "techniques" rather than "technologies". These global approaches provide a new blow to progress in favor of "technical trajectories", which are "constructed as much by political powers, economic actors and human communities" (Bray and Hilaire-Pérez 2019). They also restore to techniques the complexity and multiplicity of their geographical and temporal trajectories.

This global history of techniques resonates with what has been interpreted as a "third generation" of environmental history in Europe (Bécot 2022b). This is characterized, on the one hand, by a multiplication of fields of analysis outside the United States and, on the other hand, in the context of global warming and the Anthropocene, by a dual movement to analyze global dynamics and multiscalar approaches. Signs of the relevance and dynamism of these crossovers are the explicit inclusion of young researchers in both fields, as well as the participation of several environmental historians in the project "A Global History of Technology" or in the activities of the GDR Techniques et production dans l'histoire (GDR Techniques and production in history), on subjects such as energy, conflicts,

---

14 In France, see the collective works coedited by Hilaire-Pérez and Verna (2016) for the modern period, and by Carnino et al. (2023) for the contemporary period. In Europe, see the ERC project A Global History of Technology, 1850-2000, led by the University of Darmstadt since 2017, and Mikael Hård (2023).

15 The national *Historiographies of Techniques* conference organized by the GDR Techniques et production dans l'histoire (GDR Techniques and production in history) addresses one of the ambitions of global approaches, which is to deconstruct the national framework that has prevailed in historiographies of techniques, despite the criticisms it has received, starting with David Edgerton (1998).

pollution, waste and temporalities, or the contribution of historians of techniques such as those of the Groupe d'intérêt scientifique, GIS (Scientific Interest Group) *Histoire de la mer* in a recent publication on new perspectives in environmental history (Llinares et al. 2022). The history of energy has been particularly conducive to this type of crossover: it is no longer a question, as in Debeir et al. (2013), of writing about the development and decline of civilizations and their energy systems but rather of examining resistance and alternatives to fossil fuel and nuclear energy "trajectories"[16], the territories of development and "energy transitions" (Massard-Guilbaud and Mathis 2019), and the social implications and perspectives that accompanied the advent of the "coal civilization" (Mathis 2021).

8.4. Conclusion

Geneviève Massard-Guilbaud's (2018) assertion that it is no longer possible to write the history of energy while ignoring the global climate problem that can be applied to all human activities. History of techniques still has much to contribute. A number of avenues remain open for an environmental history of techniques, which seek to understand the multiplicity of relationships with the environment induced by techniques across territories, as well as the balance of power it has brought into play at different spatial scales. Cases of environmental injustice linked to developments or productive activities such as landfills or incinerators on an urban scale (Mittlefehlt 2018) or ex-uranium mining on a global scale (Hecht 2016) are good examples. In addition, renewed attention to the materiality of environments considerably enriches the analysis of the contingencies that surround the future of techniques, which are important for questioning a linear and teleological history. The failures of projects carried out by colonial powers demonstrate the resistance of societies and environments to certain promethean ambitions while rehabilitating the relevance of local practices and knowledge in their environmental contexts (Pessis 2013; Marrec and Plarier 2025; Plarier 2025). Finally, studies have shed light on the mechanisms that lead to abandonment or, on the contrary, the persistence, disappearance or resurgence of techniques and practices that appear to be alternatives to consumerist lifestyles and an extractivist economy, such as reuse and repair (Bernasconi et al. 2022; Letté 2023) or composting (Barles and Dufour 2021). By providing an understanding of material and social legacies, as well as of the choices and impasses of the past, these avenues provide keys to thinking about the

16 To cite only the French literature: (Lopez 2014; Jarrige and Vrignon 2020). Andreas Malm's work on resistance to the transition from hydropower to coal in 19th-century England and the logic behind it has remarkably contributed to this literature (see, in particular, Malm (2016)).

role and form that techniques could take in a more socially just and ecologically sustainable world.

8.5. References

Abrassart, C., Jarrige, F., Bourg, D. (2020). Introduction. Low-tech et enjeux écologiques – quels potentiels pour affronter les crises ? *La Pensée écologique*, 5, 1.

Aggeri, F. and Cartel, M. (2017). Le changement climatique et les entreprises : enjeux, espaces d'action, régulations internationales. *Entreprises et histoire*, 86, 6–20.

Barles, S. (2005). *L'invention des déchets urbains*. Champ Vallon, Lyon.

Barles, S. and Dufour, E., (2021). L'éviction du compostage des ordures ménagères et la fin de leur recyclage agricole en France et en Ile-de-France (1940-1990) : le rôle de la valorisation marchande et de la normalisation technoscientifique. Research Report, PIREN-Seine Phase.

Barles, S. and Guillerme, A. (1995). *L'urbanisme souterrain*. Presses Universitaires de France, Paris.

Bartolotti, F. (2020). Les autorités portuaires face aux pollutions pétrolières dans l'étang de Berre et le golfe de Fos (1945-1979). *Rives Méditerranéennes*, 61, 47–68.

Bécot, R. (2022a). L'histoire environnementale au travail. Repères pour une histoire environnementale des mondes du travail. In *Écrire l'histoire environnementale au XXIe siècle : sources, méthodes, pratiques*, Frioux, S. and Bécot, R. (eds). PUR, Rennes.

Bécot, R. (2022b). Conclusion. Une "troisième génération" ? Faire de l'histoire environnementale en Europe au début du XXIe siècle. In *Écrire l'histoire environnementale au XXIe siècle : sources, méthodes, pratiques*, Frioux, S. and Bécot, R. (eds). PUR, Rennes.

Bécot, R., Devienne, E., Fournier, P., Frioux, S., Mathis, C.-F., Rainhorn, J. (2023). *Le chemin, la rive et l'usine. Faire de l'histoire environnementale avec Geneviève Massard-Guilbaud*. Presses Universitaire des Mines, Paris.

Bernasconi, G., Carnino, G., Hilaire-Pérez, L., Raveux, O. (eds) (2022). *Les réparations dans l'histoire. Cultures techniques et savoir-faire dans la longue durée*. Presses des Mines, Paris.

Bernhardt, C. and Massard-Guilbaud, G. (2002). *La pollution dans les sociétés urbaines et industrielles d'Europe. The Modern Demon. Pollution in Urban and Industrial European Societies*. Presses universitaires Blaise Pascal, Clermont-Ferrand.

Bijker, W.E., Hughes Thomas, P., Pinch, T. (eds) (1987). *The Social Construction of Technological Systems. New Directions in the Sociology and History of Technology*. MIT Press, Cambridge.

Blanchard, A. (1979). Les ingénieurs du "roy", de Louis XIV à Louis XVI : étude du corps des fortifications. PhD Thesis, University of Montpellier, Montpellier.

Blandin, P. (1992) De l'écosystème à l'écocomplexe. *Sciences de la nature, sciences de la société : les passeurs de frontières*. CNRS Éditions, Paris.

Bonneuil, C. and Fressoz, J.-B. (2013). *L'événement anthropocène*. Le Seuil, Paris.

Bonneuil, C., Choquet, P.-L., Franta, B. (2021). Early warnings and emerging accountability: Total's responses to global warming, 1971–2021. *Global Environmental Change*, 71(4).

Bonnot, T. and Letté, M. (2023). Repair, recycling and waste. In *Global History of Techniques (19th-21st Centuries)*, Carnino, G., Hilaire-Pérez, L., Lamy, J. (eds). Brepols, Turnhout.

Booker, M. (2013). *Down by the Bay: San Francisco's History between the Tides*. University of California Press, Berkeley.

Bray, F. and Hilaire-Pérez, L. (2016). Introduction. Les techniques et l'histoire globale. In *Histoire des techniques. Mondes, sociétés, cultures (XVIe-XVIIIe siècle)*, Hilaire-Pérez, L., Carnino, G., Kobiljski, A. (eds). PUF, Paris.

Carnino, G., Hilaire-Pérez, L., Lamy, J. (2023). *Global History of Techniques (19th-21st Centuries)*. Brepols, Turnhout.

Caron, F. (1987). L'innovation et l'histoire. *Histoire, économie et société*, 2, 149–154.

Caron, F. (1998). *Les deux révolutions industrielles du XXe siècle*. Albin Michel, Paris.

Chatterji, O. (2022). La "vallée de l'aluminium", une histoire environnementale des territoires alpins face aux pollutions de la grande industrie. PhD Thesis, Université Grenoble-Alpes, Grenoble.

Cronon, W. (1991). *Nature's Metropolis, Chicago and the Great West*. W.W. Norton & Company, New York.

Daumas, M. (1979). *Histoire générale des techniques*. PUF, Paris.

Daumas, J.-C. and Mioche, P. (2004). Histoire des entreprises et environnement : une frontière pour la recherche. *Entreprises et histoire*, 35(1), 69–88.

Debeir, J.-C., Deléage, J.-P., Hémery, D. (2013). *Une histoire de l'énergie. Les servitudes de la puissance*. Flammarion, Paris.

Demuth, B. (2019). *Floating Coasts. An Environmental History of the Bering Strait*. W.W. Norton & Company, New York.

Febvre L (1935). Réflexions sur l'histoire des techniques. *Annales d'histoire économique et sociale*. 7(36), 531–535.

Fernandez, M. and Guillerme, A. (2018). L'histoire des techniques est aussi une histoire environnementale. [Online]. Available at: http://journals.openedition.org/ephaistos/1160 [Accessed 11 August 2023].

Fournier P. (2003). Aménagements hydrauliques et structuration de l'espace : les métamorphoses de l'eau en Provence et Comtat. *Dix-septième siècle*, 221(4), 585–601.

Fournier, P. (2005). Une histoire des techniques pour quoi faire ? Quelques orientations pour la période moderne. *Siècles*, 22, 95–107.

Fournier, P. and Massard-Guilbaud, G. (eds) (2016). *Aménagement et environnement. Perspectives historiques*. PUR, Rennes.

Fressoz, J.-B. (2012). *L'apocalypse joyeuse. Une histoire du risque technologique*. Le Seuil, Paris.

Fressoz, J.-B. and Jarrige, F. (eds) (2013). L'histoire et l'idéologie productiviste. Les récits de la "révolution industrielle" après 1945. In *Une autre histoire des "Trente Glorieuses". Modernisation, contestations et pollutions dans la France d'aprèsguerre*. La Découverte, Paris.

Fressoz, J.-B., Graber, F., Locher, F., Quenet, G. (2014). *Introduction à l'histoire environnementale*. La Découverte, Paris.

Garçon, A.-F. (2012). *L'imaginaire et la pensée technique*. Classiques Garnier, Paris.

Gardner, R. (2009). Constructing a technological forest: Nature, culture, and tree-planting in the Nebraska Sand Hills. *Environmental History*, 14, 275–297.

Gille, B. (1978). *Histoire des techniques*. Gallimard, Paris.

Gras, A. (1998). Anthropologie et philosophie des techniques : le passé d'une illusion. *Socio-anthropologie*, 3.

Griset, P. and Bouvier, Y. (2012). De l'histoire des techniques à l'histoire de l'innovation. Tendances de la recherche française en histoire contemporaine. *Histoire, économie et société*, 31(2), 29–43.

Guillerme, A. (1983). *Les temps de l'eau : la cité, l'eau, les techniques*. Champ Vallon, Seyssel.

Guillerme, A. (2007). Éditorial. *Documents pour l'histoire des techniques*, 14, 3–4.

Guillerme, A. and Barles, S. (1995). *L'urbanisme souterrain*. PUF, Paris.

Hård, M. (2023). *Micro-Histories of Technology. Making the World*. Palgrave Macmillan, Gewerbestrasse.

Hatzfeld, N., Pigenet, M., Vigna, X. (eds) (2016). *Travail, travailleurs et ouvriers d'Europe au XXe siècle*. Éditions universitaires de Dijon, Dijon.

Hecht, G. (2014). *Le rayonnement de la France. Énergie nucléaire et identité nationale après la Seconde Guerre mondiale*. Éditions Amsterdam, Paris.

Hecht, G. (2016). *Uranium africain. Une histoire globale*. Le Seuil, Paris.

Hilaire-Pérez, L. (2019). L'histoire des techniques a longtemps été la discipline la plus simplificatrice. Entretien avec Liliane Hilaire-Pérez. *Zilsel*, 5(1), 229–267.

Hilaire-Pérez, L. and Verna, C. (2016). Histoire économique et histoire des techniques (XVe-XVIIIe siècle). *Artefact*, 4, 13–30.

Hilaire-Pérez, L. Carnino, G., Kobiljski, A. (eds) (2016). *Histoire des techniques. Mondes, sociétés, cultures (XVIe-XVIIIe siècle)*. PUF, Paris.

Hughes, T.P. (1989). The evolution of large technological systems. In *The Social Construction of Technological Systems. New Directions in the Sociology and History of Technology*, Bijker, W.E., Hughes Thomas, P., Pinch, T. (eds). MIT Press, Cambridge.

Hughes, T.P. (2004). *Human Built World: How to Think About Technology and Culture*. University of Chicago Press, Chicago.

Jarrige, F. (2009). *Au temps des "tueuses de bras". Les bris de machines à l'aube de l'ère industrielle (1780-1860)*. PUR, Rennes.

Jarrige, F. (2014). *Technocritiques. Du refus des machines à la contestation des technosciences*. La Découverte, Paris.

Jarrige, F. (2017). L'historien et la question écologique. *Histoire@Politique*, 31(1), 75–83.

Jarrige, F. and Vrignon, A. (2020). *Face à la puissance. Une histoire des énergies alternatives à l'âge industriel*. La Découverte, Paris.

Jas, N. and Boudia, S. (eds) (2013). *Toxicants, Health and Regulation Since 1945*. Pickerng & Chatto, London.

Jones, G. (2019). *Profits and Sustainability. A History of Green Entrepreneurship*. Oxford University Press, Oxford.

Larrère, C. (2010). Les éthiques environnementales. *Natures Sciences Sociétés*, 18(4), 405–413.

Larroque, D. (1991). Le Centre d'histoire des techniques (CDHT) du Conservatoire national des Arts et Métiers (CNAM). *Bulletin de l'Institut d'Histoire du Temps Présent*, 45, 21–25.

Latour, B. (1993). *La clé de Berlin et autres leçons d'un amateur de sciences*. La Découverte, Paris.

LeCain, T. (2005). What is envirotech? *The Envirotech Newsletter*, 1(1), 4–5.

LeCain, T. (2009). *Mass Destruction: The Men and Giant Mines that Wired America and Scarred the Planet*. Rutgers University Press, New Jersey.

Llinares, S., Tranchant, M., Sauzeau, T. (2022). Ports et environnement, une histoire exemplaire à écrire dans la longue durée. In *Écrire l'histoire environnementale au XXIe siècle : sources, méthodes, pratiques*, Frioux, S. and Bécot, R. (eds). PUR, Rennes.

Lopez, F. (2014). *Le rêve d'une déconnexion. De la maison autonome à la cité autoénergétique*. La Villette, Paris.

Mackenzie, D. and Wajcman, J. (eds) (1999). *The Social Shaping of Technology. How the Refrigerator Got its Hum*. Open University Press, Milton Keynes/Philadelphia.

Malm, A. (2016). *Fossil Capital: The Rise of Steam Power and the Roots of Global Warming*. Verso Books, London.

Marrec, A. (2018). Histoire des énergies renouvelables en France, 1880-1990. PhD Thesis, Université de Nantes, Nantes.

Marrec, A. (2022). Wave converters against energy systems? Le Minou's Ram and the Cattaneo Ondo-Pump in the frame of French Blue Coal Policy (1925-1945). *Cahiers François Viète*, 3(12), 49–72.

Marty, N. and Druelle-Korn, C. (2023). Histoire des entreprises et économie circulaire : pour une nécessaire clarification. *Entreprises et histoire*, 1, 6–17.

Massard-Guilbaud, G. (2002). De la "Part Du Milieu" à l'histoire de l'environnement. *Le Mouvement Social*, 200, 64–72.

Massard-Guilbaud, G. (2007). Pour une histoire environnementale de l'urbain. *Histoire urbaine*, 18(1), 5–21.

Massard-Guilbaud, G. (2010). *Histoire de la pollution industrielle*. EHESS, Paris.

Massard-Guilbaud, G. (2018). Introduction. De l'histoire des filières à l'histoire des systèmes énergétiques. In *Systèmes et transitions du Moyen Âge à nos jours*, Massard-Guilbaud, G. and Mathis, C.-F. (eds). Presses Universitaires de la Sorbonne, Paris.

Mauss, M. (1934). Les techniques du corps [Online]. Available at: https://anthropomada.com/bibliotheque/Marcel-MAUSS-le-corps.pdf [Accessed 5 September 2025].

Melosi, M.V. (1994). The place of the city in environmental history. *Environmental History Review*, 1(7), 1–23.

Mumford, L. (2016). *Techniques et civilisations*. Parenthèses, Marseille.

Parrinello, G. (2018a). Systems of power: A spatial envirotechnical approach to water power and industrialization in the Po Valley of Italy, ca.1880-1970. *Technology and Culture*, 59(3), 652–688.

Parrinello, G. (2018b). L'environnement de l'histoire environnementale : un enjeu politique ? [Online]. Available at: http://journals.openedition.org/histoirepolitique/9080 [Accessed 5 September 2025].

Parrinello, G., Bécot, R., Caligari, M., Yrygoy, I. (2022). Shifting shores of the anthropocene: The settlement and (unstable) stabilisation of the North-Western Mediterranean littoral over the nineteenth and twentieth centuries. *Environment and History*, 28(1), 129–154.

Pérez, L. (2009). L'histoire intellectuelle des techniques au Centre d'histoire des techniques et de l'environnement du Conservatoire national des arts et métiers (CDHTE-CNAM). *Revue de synthèse*, 130, 147–164.

Pessis, C. (2013). La machine au secours de l'empire colonial ? La mécanisation de l'agriculture et ses détracteurs en Afrique tropicale française. In *Une autre histoire des Trente glorieuses*, Pessis, C., Topçu, S., Bonneuil, C. (eds). La Découverte, Paris.

Pestre, D. and Cohen, Y. (1998). Présentation. *Annales. Histoire, sciences sociales*, 53(4/5), 721–744.

Picon, A. (1988). *Architectes et ingénieurs au siècle des Lumières*. Parenthèses, Marseille.

Picon, A. (1992). *L'invention de l'ingénieur moderne. L'École des ponts et chaussées, 1747-1851*. Presses de l'École nationale des ponts et chaussées, Paris.

Picon, A. (1995). Construction sociale et histoire des techniques. *Annales. Histoire, Sciences Sociales*, 50(3), 531–535.

Plarier, A. (2025). "Territorialiser l'empire. Un nouvel environnement colonial en république". In *Les natures de la République. Une histoire environnementale de la France, 1870-1940 (Vol. 2)*, Cornu, P., Frioux, S., Marrec, A., Mathis, C.-F., Plarier, A (eds). La Découverte, Paris.

Plarier, A. and Marrec, A. (2025). "L'empire au temps de sa mise en valeur, l'illusion d'un apogée". In *Les natures de la République. Une histoire environnementale de la France, 1870-1940 (Vol. 2)*, Cornu, P., Frioux, S., Marrec, A., Mathis, C.-F., Plarier, A (eds). La Découverte, Paris.

Pritchard, S.B. (2011). *Confluence. The Nature of Technology and the Remaking of the Rhône*. Harvard Historical Studies, Cambridge.

Pritchard, S.B. (2014). *Toward an Environmental History of Technology*. The Oxford Handbook of Environmental History, Oxford.

Rainhorn, J. (2019). *Blanc de plomb. Histoire d'un poison légal*. Les Presses de Sciences Po, Paris.

Rand, L.R. (2017). Beyond the biosphere: Expanding the limits of the human world. In *Living in the Anthropocene: Earth in the Age of Humans*, Kress, W.J. and Stine, J.K. (eds). Smithsonian Institution, Washington, D.C.

Reuss, M. and Cutcliffe, H. (2010). *The Illusory Boundary. Environment and Technology in History*. University of Virginia Press, Charlottesville.

Russell, E., Allison, J., Finger, T., Brown, J.K., Balogh, B., Bernard Carlson, W. (2011). The nature of power: Synthesizing the history of technology and environmental history. *Technology and Culture*, 52(2), 246–259.

Salomon, J.-J. (1980). Daumas Maurice, Histoire générale des techniques. Les techniques de la civilisation industrielle. Gille Bertrand, Histoire des techniques. Technique et civilisations. Technique et sciences. *Revue française de sociologie*, 21(3), 455–461.

Stine, J.K. and Tarr, J.A. (1998). At the intersection of histories: Technology and the environment. *Technology and Culture*, 39(4), 601–640.

Van der Straeten, J. (2019). Borderlands of industrial modernity: Explorations into the history of technology in Central Asia, 1850-2000. *Technology and Culture*, 60(3), 659–687.

Van der Straeten, J. and Weber, H. (2025). Technology and its temporalities. A global perspective. In *Global History of Techniques (Nineteenth to Twenty-First Centuries)*, Carnino, G., Hilaire-Perez, L., Lamy, J. (eds). Brepols, Turnhout.

Vérin, H. (1993). *La gloire des ingénieurs. L'intelligence technique du* XVIe *au* XVIIIe *siècle*. Albin Michel, Paris.

White, R. (1995). *The Organic Machine: The Remaking of the Columbia River*. Hill and Wang, New York.

Woronoff, D. (ed.) (1990). *Forges et forêts: recherches sur la consommation protoindustrielle de bois*. Éditions de l'École des hautes études en sciences sociales, Paris.

9

Techniques and Gender

Delphine GARDEY
Université de Genève, Geneva, Switzerland

9.1. Women, machines and progress

With respect to the history of gender and techniques, Michelle Perrot's seminal article, "Femmes et machines au XIXe siècle" (Women and machines in the 19th century), published in 1983, is still a source of inspiration and rejuvenation (Perrot 1983). It was a time of reinterpretation by Marxism. Historians and sociologists alike have continued to focus on industrial work. With respect to techniques, the aim was to understand the role they played in the production process and how they were used in capital/labor conflicts to control labor or maximize profit (Braverman 1974; Noble 1984). All we know of labor is masculine, necessarily "universal". As a figure of present exploitation or future emancipation, the worker dominates the narratives. For the most part, the focus is on industrial machinery. Ultimately, a deterministic conception prevails: the aim is to measure the "impact" of techniques, their "effects", or to indicate how they "reflect" capitalism. Techniques are then used to support or oppose "progress".

Michelle Perrot shifts many of these expectations. She gives women a key place, reminding us that they are ordinary workers, not occasional extras on the conventional stages of industrial history. Women's work is always there, in the domestic sphere as a space of production, as in the workshop. The challenge is to study the way in which the division of labor and the recomposition of tasks operate in any given sector, place and technique, whether they are mechanical or not. It is also about highlighting women's technical skills. Are women "promoted" by machines? Yes and no, she answers. If they are, it is more likely to be at the end of

History of Techniques,
coordinated by Guillaume CARNINO and Xavier GUCHET.

the 18th century than during the 19th century. Thus, she continues, drawing on Hélène Robert's dissertation (1980), "the first phase of textile mechanization, that of spinning, saw the rapid disappearance of spinning wheels… With the advent of the new English spinning machines, work was reorganized, with the head of the family, who became the team leader, the beneficiary; he employed his children, and his wife, as auxiliaries, the father as foreman, and for this he received an overall family salary […]. In the indienage[1], the introduction of the mechanical roller around 1836–1840 eliminated the need for women and children, and reinforced the masculinity of the trade". She concludes: "As soon as a machine is deemed complex (and this complexity needs to be assessed), men retain control of it". While it was repeatedly asserted that men's muscular strength and women's "congenital weakness" justified inequalities in status and pay, the introduction of the jenny mule did not benefit women, despite its potential to automate and lighten the workload of spinning, but instead elevated spinners. It is therefore clear that "mechanization does not have clear effects" because "the place of women is not regulated by techniques, but by questions of status" (Perrot 1983). In other words, the relationship with techniques is already a social and gender-based relationship (Chabaud-Rychter and Gardey 2002). While this relationship – and this is undoubtedly one of the strengths of historical analysis – is local, contextual, situated and subject to transformation, there are a number of reiterations and regularities in the trifecta formed by the relations among men, women and techniques in the history and anthropology of activities and work.

9.2. Tools, appropriation and qualification

Michelle Perrot refers to the work of anthropologist Tabet (1979), who undertook a meta-analysis of literature to describe the situation of "hands, tools and weapons" from the prehistory of human societies to societies studied by 20th-century anthropologists. Structurally underequipped, women are said to be technically subordinate. The control of tools and weapons became a male privilege. In hunter–gatherer societies, women had to make do with rudimentary tools, whether for archaic or more complex tasks. Equipped in a way that limited their gains in fishing and hunting, they would have been limited in their access to meat and protein throughout evolutionary history, as Touraille (2008) later set out to demonstrate. In harvesting and agricultural societies, they are underequipped for these activities. Conversely, tools that are strictly reserved for men, when they are not strategic to agricultural production, enable the manufacture of other tools. As Tabet (1979) writes, with "technological evolution", "the human being is no longer defined and limited by the possibilities of his body", with "tools" becoming "his extension" and expanding "his capacity to appropriate nature and act upon it"

1 Printed cotton fabrics industry.

(p. 11). In that context, women find themselves paradoxically limited to their own bodies, to bare-hand operations and rudimentary tools. Furthermore, they are used as bodies that can be appropriated.

These readings take up the question of the appropriation and expropriation of time, knowledge, techniques and women's bodies in the course of industrialization. We are thinking here of the long, dominant period in the 17th and 18th centuries, when industrial production was organized by subcontracting (*Verlagsystem* or *Putting out*). The craftsman-head of the family ran the domestic unit that formed the basic economic and technical unit. He controlled the process and generally owned the tools but would not control the raw materials or the final product. He depended entirely on the client, who remunerated the head of the family, not for the productive time of all his members but for the finished product, whose value and outlets he controlled. Veltz (2000) speaks of this unit as a place where "agricultural, industrial and domestic tasks are balanced" (p. 49). A gentle euphemism that conceals the appropriation and expropriation of women's reproductive and productive work within the family, as, later on in the wage economy: with the double working day, the half-wage, the legal guardianship of the wife and mother. This "patriarchal" vein that accompanies the industrial development of capitalist economies has been theorized by Delphy (1998) in her "criticism of political economy" and her characterization of the monopolization of women's productive and reproductive labor by the institution of marriage, as more generally in society.

Finally, Michelle Perrot examines the specific forms taken by the relationship with technique in the factory system as a closed, disciplinary, hierarchical and organized space. For men, it is the authority and complex tools that redeploy their ability to act on the world and others. For women, it is the subordinate position, the assignment to body and hands, the surveillance and limitation of movement, and the time-consuming work. The initial idea that women were destined for fragmented, repetitive, low-skilled and technically limited tasks was born with industrialization. Contemporary work by German historian Karin Hausen (2012) shows evidence of this, demonstrating how "the economic logic of cost reduction sometimes bowed to the logic inherent in the so-called natural gender order". Pointing out how the "gender hierarchy" functions as an "economic resource", she and others (Tilly and Scott 2002) suggest that a history of industrialization cannot be written in abstraction from gender relations. This is because the latter affect the whole of technical and capitalist development and are, in turn, affected by it.

The question then revolves around the technical and professional non-recognition of women's tasks, which is repeated from one industrialization to the next, from one technical innovation to the next. Women's qualifications are not recognized, nor are the techniques they master. Historical and sociological research on qualifications, while it has focused on techniques, has persistently overlooked gender relations. Feminist historians and sociologists were the first to highlight the constant social and gendered nature of what is defined as "qualification" at work and in employment (Perrot 1978; Kergoat 1982; Maruani and Nicole 1991; Gardey 1998, 2013; Daune-Richard 2003). More generally, Wajcman (1991) pointed to the existence of a vicious circle: "women's lack of technical competence actually becomes part of the female gender identity, at the same time as a gender stereotype".

9.3. Machinism, techniques and masculinities

We could say, following the remarks made by Cockburn (1991, 2004) and Wajcman (1991, 2004), who were pioneers in the study of techniques, that techniques have historically been defined in the West as what men do. Wajcman (2002) summarized this as follows: "The conventional definition of technique is always presented more or less in terms of male activities. We tend to think of technique in terms of industrial machinery or cars, for example, rather than cooking appliances or baby strollers" (p. 303).

These points have been taken up in very different works, such as those by Berner (1997) on the culture of engineering schools or by Turkle (1988) on computer hackers. They describe social and professional situations almost a century apart but reveal certain constants: the production of a masculine inner space, socialization via self-testing and the advent of machines. In particular, Berner shows how technical culture appears to be one of the constitutive elements of male identity. As a result, it appears that women are excluded from professions or places where they "do" techniques, no doubt because techniques define masculinity.

Ruth Oldenziel explored this question in greater depth, looking at the production of "masculinities" in industrialized societies. She studies the role played by the toy industry in the relationship between techniques and young Americans in the early 20th century. The institutionalized production of a consumer market for boys' toys (toy cars, miniature trains and circuits, bicycles, construction sets, model airplanes and ships) helped shape their social identity, as well as the definition of their professional future, through an anticipated, expert and legitimized relationship with the leading techniques of the time (Oldenziel 1997). She also shows how "techniques" and "inventions" are redefined in the 19th century and more strictly identified with machinism, thus "objecting" to the naturalness of this construction, which is, above all, historical and social (Oldenziel 1996). As a product of the industrial economy, as a means of producing new objects, machines asserted themselves as the "engine" of capitalism in the 19th and 20th centuries. Machinism and the production of goods were firmly established as the center and promise of progress, as well as the technical and masculine sphere (Horowitz 2001). In *Making Technology Masculine*, Oldenziel (1999) traces the historical process by which men claimed technology as their "exclusive terrain". From the workshop to the world's fairs, from workers, technicians, engineers and inventors to trade unionists, activists and novelists, she stages and questions technology as a "modern masculine myth". It is as much a question of intellectual construction and discourse as material practices and real-life social scenes. "Technology" establishes itself as the narrative of a modern form of virility.

9.4. Technical cultures and women's industries

Put another way by the sociologist of techniques Madeleine Akrich, we need to reverse the chains of allocation: it is not because women are technically incompetent that technological choices and the "positions" associated with certain activities are in the hands of men, but it is because technology has been actively constructed by men as a mechanism of exclusion that women appear incompetent. "What we call techniques", she writes, "is indeed a gender construction" (Akrich 2002, p. 96).

Mining, metal and "hard" industries, the quintessential (but not exclusive) masculine sectors, have long overshadowed the initial and decisive nature of the cotton, textile, silk, processing, food and tobacco industries, as well as the many places and activities in which women have contributed to economic development. Like American historian Judy McGaw (1982, 2003), a pioneer in the history of techniques and gender, we can affirm that "women's technologies count". She asks us, "What is technical, and what is not?" What is technical in material cultures? The answers are highly contingent. Such is the case with needlework, which punctuated the hours of most women's days in the 19th and early 20th centuries, which is both an essential activity of self-production and repair, of work at home for oneself or for others, and a salaried activity for women who perform it in workshops. The techniques and professional skills involved refer to a wide range of trades and sectors: lingerie, corsetry, hosiery, hand or mechanized embroidery, millinery, trimmings, flowers and feathers. For working-class women, sewing was a necessity and a means of earning a living; for young girls and bourgeois women, it was a leisure activity and an ostentatious sign of idleness. Finally, sewing is a structuring element in the production of gender identity, in the socialization of women to a social role, an essential place for the definition of a feminine culture of techniques as much as a feminine "technique" (Zylberberg-Hocquard 2002).

These points have been taken up and reworked by Auslander (2014) from the perspective of the history of "gendered textile manufacture". The approach in terms of "material culture" (Kingery and Lubar 1995) suggests a less restrictive definition of techniques, potentially more inclusive of "women's work" (Perrot 1978). Textiles have been produced from antiquity to the present day "in town and country, in monasteries, guilds, domestic workshops or factories" (Auslander 2014). For the most part, this history is told in feminine terms, even if men have performed and still perform some of these activities (whether or not they are mechanized), whether they are the tailors of the Old Regime or the sewers of leather and skins in Europe – which have always been masculine – or the past and present neighborhood tailors in Africa.

Strangely, the invisibility of sewing continues with mechanization. The first mass-produced mechanical object in the history of the world, the sewing machine, has not been given the credit it deserves. This essential instrument in the history of capitalism (as a production good as well as a consumer good) has probably suffered from its "feminine" roots. Initially introduced as a production tool in the workshop, mechanizing the work of seamstresses, the sewing machine enjoyed considerable success "in the bedroom", being purchased on a massive scale by those who sewed at home. By 1890, 90% of French households were equipped with the instrument (Jarrige 2018). This development satisfied the "philanthropic" demand for a female worker "protected" from the harmful effects of the factory and the laws of the market because she was "at home" (Peyrière 1992). As an exploitation tool and a renewed form of the sweating system, the sewing machine does not credit the productive and technical contributions of women. It is when we look at the rise of consumerism that we realize its importance. The sewing machine, commercialized by the new technologies of mail order and credit purchasing, along with clothing, one of the major sectors of mass consumption, was indeed a decisive tool in capitalist transformations between the two centuries (Coffin 1996).

9.5. Invent, innovate and undertake

Women's technical skills and their role in the economic and technological development of Western societies have been profoundly neglected. This is particularly true in the field of invention and innovation. The challenge is to question these notions from a gender perspective. From this perspective, historiography (essentially North American) seems to be more advanced in the "history of science" than in the "history of techniques" – even if the distinction between these two fields remains problematic and has given rise to numerous discussions and contextualizations. Aware of the famous "Matthew/Matilda Effect" in science – the fact that we only lend to the rich and that women are rarely credited with decisive advances, even if they contribute to the development of science and technology (Rossiter 2003) – some have set out to rewrite the history of techniques to search for the contributions of "mothers" and "daughters of invention" in rarely frequented historical scenes, archives and spaces (Stanley 1993). While we need to account for the historically situated expectations of the notion of invention (Hilaire-Pérez 2000), there is also the challenge of reprogramming the dominant narratives on women's contribution to techniques, from invention to use, by broadening the spectrum of scenes and contexts in which we can and must look (Schibienger and Schraudner 2011). This is a question of rediscovering the "inventive" sectors (Chanteux 2009), which, because they are feminized, have been discredited or, because they are masculine, have rendered those who invent and innovate invisible. The engineering worlds have undergone a profound reinterpretation in this respect, with numerous internationally comparable initiatives on pioneering women in the fields of engineering science and technology (Pursell 1993; Canel and Zachmann 1997; Bix 2014). Women's entrepreneurship should still be investigated in terms of its dominant technological variants and cultures, as well as from the point of view of feminine socialized fields such as social assistance, medical treatment and, more generally, care, whose technologies have yet to be defined.

9.6. Consume, choose and mobilize

Noting that women have been underrepresented in the legitimized fields of technique and that historians of technology have often focused on the design and production of technology, feminist historians and sociologists have suggested that by focusing on users and use, rather than engineers and design, it might be possible to write histories other than those of men inventing and mastering technology (Wajcman 1991; Lehrman et al. 1997).

In the 1990s, a new historiography focused on the history of management, marketing (Cochoy 1999), advertising and commercial mediation and highlighted the role of consumer leagues in the development of certain industries. It is as much a question of bringing management tools (Yates 1989; Gardey 2008) into the history of techniques as it accounts for the economic and technical history of consumption (Strasser 1995), possibly from a gender perspective (Horowitz and Mohun 1998). In her history of home and domestic technologies, Cowan (1983) quips about the meaning of progress (household appliances ultimately meaning "more work for women") and focuses on consumption as a place where different technologies compete and as a possible space for arbitration. The term *consumption turn* has been used to describe the shift in perspective toward market mediation, users and commodification (Oudshoorn and Pinch 2003).

Historians (especially in the English-speaking world) are addressing the indissociably technical and cultural aspects of consumption. Beyond objects and artifacts, attention is being given to the renewed forms of trade and sale and to the material and immaterial techniques that support them. The question of the historicity of the so-called spheres of production and consumption has also been reopened, as has the analysis of practices of distinction and leisure over the long history of social classes. Questioning the "sex of things" (de Grazia 1996) makes it possible to revisit the commonplaces of gendered definitions of these spheres (production for men, consumption for women), to qualify different "regimes of consumption" and to look at how consumption manifests and defines social and gender identities (Auslander 1996). Women stand out not only as producers of goods and commodities but also as "citizen-consumers", with, for example, the long tradition of boycotting imported products from the United Kingdom to protect the national identity of revolutionary America (Auslander 2010). Attempting another history of Europe, Oldenziel and Hård (2013) focus on the popularization of ordinary or "low-tech" systems, which are made up of "transnational user movements" in which consumers, activists and handymen play an important role, helping to shape "forgotten scenes of European history".

9.7. Interpret, interact and socialize

This work within the historical field is linked to broader transformations in the so-called "social study of science and technology" that precede and accompany their development. In sociology, the SCOT (Social Construction of Technology) movement (Bijker et al. 1987) seeks to account for the "non-finite" nature of technologies, basing its work on detailed analysis of the early history of certain techniques (such as the bicycle) or materials (such as bakelite) (Bijker 1995). The aim is to show how, beyond the scripts of the inventors, these techniques are socialized and redefined through use. This approach speaks of the "interpretative flexibility" and "stabilization" of techniques, notions that have been the subject of subsequent discussions and contributions. In the French sociological tradition of innovation (Akrich et al. 2006), we look at how technologies are coconstituted by actors: the idea here is that innovation indissociably produces artifacts and social bonds, techniques and society. In these arrangements, technical devices play a major role. Technical choices thus appear as social processes, in the broadest sense (Akrich 2002). Ultimately, the renewal brought about by the social study of science and technology in its various traditions and paradigms (Pestre 1995; Cohen and Pestre 1998) has strongly challenged the classical vision of users as passive consumers of technology, as well as the linear model of technological innovation and diffusion (Edgerton 1998, 2007). The analysis of sociotechnical controversies, in various forms, has also nurtured a non-deterministic interpretation of the development of science and technology in society, making room for the variety of scenarios and scripts that exist in terms of the technical response to a question or the technical (and thus social and political) definition of a problem at a given moment.

Placed at the heart of the analysis, the field of technical objects and practices that are likely to give rise to analysis marks, in any case, a disruption (MacKenzie and Wajcman 1985), testifying to a kind of "technical" turning point in the social sciences. However, what is the place given to gender issues in these different traditions?

9.8. Scripts, facts and identities

The focus on users also reflects an interest in objects and everyday life, which is a broadening of the sphere of what can be defined as technical and a new reflection on the links between techniques and society or on the place of the body in technical mediation. Feminist critique of techniques (and science) plays a decisive role in bringing about new objects and new questions that directly touch on issues of gender, bodies and sexuality (Creager et al. 2001). More fundamentally, as the period was more "agnostic", detaching itself from structuralism and Marxism, this critique proposed less deterministic but also less fixed conceptions of techniques and gender. As Judy Wajcman writes, "feminist studies of gender and techniques usually theorize gender as a fixed, unambiguous phenomenon, existing prior to and independent of techniques". However, "the aim is to account for the social shaping of techniques", not to define gender and/or techniques as a given but rather as conceived and redefined in social interaction (Wajcman 2002, p. 63).

A famous example of this type of approach is the investigation into the introduction of the microwave oven by Cockburn and Omrod (1993). A domestic adaptation of military technology, the microwave was initially imagined as a "brown product" (as was the VCR in its day), aimed at male executives, unlike the "white products" fashioned and marketed for women (the "queens of the home"). However, this initial script failed to meet with consumers, forcing the object to be disenfranchised and regenerated to appeal to those who do the cooking on a daily basis. Our focus on innovators and our detailed analysis of the product's design and marketing reveal the ways in which technical objects are socialized in use, which are defined for potential users who are themselves subject to social and gender qualification. The social shaping of techniques and gender is therefore reciprocal.

Gender can thus be defined as "doing gender" (West and Zimmerman 1987) or as "performance", play, interpretation or reiteration (Butler 2006). It is made with techniques, just as techniques make (and are made with) gender relations. This turning point opens up a number of perspectives for denaturalizing both techniques (and their social and gendered properties) and gender (and the situated and contingent nature of social positions and roles). In this line of thought, we have studied the ways in which the machines and administrative tools of the early 20th century acquire a "gender" and how certain technical and social properties are adopted, stabilized and ultimately naturalized – in other words, how "objects and humans in action" contribute to the "commodification of male domination" but also how technical and social innovation can go hand in hand with the commodification and reinforcement of social and gender inequalities (Gardey 2002).

The fact that technical objects can catch "a sex" (in fact, more likely a "gender") invites us to consider the plasticity and flexibility of techniques, as much as those of sex/gender identities. The convergence between these so-called "constructivist" theories and contemporary changes in body states, sex/gender and sexuality, which they both aim to qualify and help produce, is not insignificant – whether we are talking about the mediations and redeployments brought about with and by so-called information and communication technologies or the transformations brought about by the rise of biomedical technologies.

9.9. Communication, information technology and digital worlds

From the telephone to the computer, the history of the 20th century teaches us that the use of techniques extends far beyond the workplace into all aspects of personal life, leisure and intimacy. In this sense, tangible and intangible technologies are not just artifacts, but also cultures, and this "cultural" aspect of techniques takes on even greater significance, as these societies and economies have become increasingly "communicational", "informational" and "digital". The boundaries between the physical and the immaterial, the soft and the hard in technological matters, have been profoundly reworked over the course of the 20th and 21st centuries (Edwards 2003). The hybridity, physicality and fluidity attributed to many of these technical artifacts and systems are still open to question. One of the challenges of the history of gender and techniques is to contribute to a more realistic history of women's contribution to the communication, computer and digital economies and worlds. As key actors in the first informational and managerial revolution of the late 19th and early 20th centuries (Gardey 2002) and as privileged users of machines for processing and producing written and encrypted information throughout the 20th century, women have also played a decisive role in the history of computing since the end of the Second World War. However, this reality is largely ignored and underestimated (Gardey 2015). Here, again, we see how the way in which we account for the history of techniques is contingent. For a long time, the history of computing was written without considering mechanography as a technical and social system capable of processing and producing data on a large scale (Gardey 2008). Computer science's place in the genealogy of cutting-edge scientific fields, that of concepts and formalisms, prevailed. It is a way of writing history that is also a manifestation of what "technique" is. The features of this internalism or technological naturalism are on the "hard" side, the "masculine" side, the innovative side, which is radically new, as feminist theorists have highlighted to characterize what techniques are, both socially and culturally, in the Western context. However, the new information-processing spaces – these mixtures of human and non-human – have relied on the intensive and specialized use of female labor, whether we are talking about the history of the telephone industry (Lipartito 1991), typing pools or mechanical accounting services, for example in the banking and insurance sectors. The history of these infrastructures must continue to be invested in (Shaffer and Thierry 2015), beyond the time of electronic accounting, after the Second World War, in those decisive years of extension of these assemblages that are the 1950s–1960s (Morley and McDonnell 2015). A remarkable feature of the early computer era is the way in which techniques themselves became mixtures of humans and machines. This is shown by the work of Light (1999) and Abbat (2012). To ask the question "Who is the computer? The machine or the woman who uses it?" is the most sarcastic way of grasping what is at stake (Light 1999). Even today, gender plays a part in defining what "software" or "hardware" is in the computational field (Lie 2003). This enables Wajcman (2006) to point out that, in the cultural process by which technologies (in this case, information technologies) are recognized as such, "gender" is the marker of what makes it possible to negotiate the "low tech/high tech" or "tech/no tech" boundaries. This gendered attribution is now dated due to the investment of young male engineers in "low tech". What is socially defined as value, skill or knowledge follows from this.

9.10. Self-techniques, sex/gender and sexualities

While we need to restore a place for corporeality in contemporary economic and social assemblages, we also need to rethink historical forms of self-experience and relationships with others, in a context where personal computers (Turkle 1995), laptops, and smartphones stand halfway between the inanimate and the living, as an extension of the self and part of the external world, as instruments and tools as much as "relational" artifacts. In the late 1980s, Donna Haraway's prophetic *Cyborg Manifesto* described the transformations underway in California in the fields of work, computing and biology (Haraway 2007). Haraway appropriates and redeploys the figure of the cyborg to consider the potentiality of a subject that escapes natural determinations and finds itself endowed with new capacities for action at both the individual and collective levels (Gardey 2016). The cyborg is proposed as an aspect beyond the body and the heterosexual regulation of reproduction. In regard to human gestation and reproduction, previously impassable boundaries are abolished. At the time she wrote her manifesto, cloning only existed in science-fiction novels[2], but the first test-tube babies had been born. Opportunistically, in this displacement of the so-called "natural" limits and capacities of incorporation, Haraway sees new possibilities for being in the world and transforming it. The cyborg becomes a means of questioning the future of the body, sex and gender in an overequipped environment. Cyberbodies bear witness to the biotechnological recomposition of body/artifact boundaries (microsurgery, implants, prostheses, repair), as well as to the more structural transformations of the ecosystem that surrounds and conditions them (medicalization, enhancement, digital mediations, etc.). The topicality of the "cyborg" issue thus became fully apparent at a time when life itself had become the object of engineering (Franklin and Ragoné 1998; Rose 2006). The living, its limits and frontiers are at the heart of contemporary technical explorations. Haraway proposed new narrative strategies (a new historical narrative) to account for these decisive transformations in the fields of biomedicine, reproduction, identity and sexuality. Feminist historians and critics of science and techniques have explored these transformations, producing decisive work, for example on the history of endocrinology (Oudshoorn 1994), the contraceptive pill (Marks 2001), the male non-pill (Oudshoorn 2003), medically assisted reproduction, sex/gender reassignment medicine and gender transition (Hausman 1995), cosmetic surgery, masculine medicine and technology (Bajeux 2022), etc.

9.11. Detour, criticism and protest

In many ways, feminist history and critique of techniques have helped to destabilize, displace, divert and subvert dominant conceptions and narratives about the social, industrial and technical world, as well as individual and collective experiences of the self and relationships to others and to these worlds. Whole swathes of our understanding of these phenomena are dependent on this work and research, which – as this rapid overview shows – has profoundly evolved in its objects of choice, as well as in its modes of interpretation and paradigms since the early 1980s and Michelle Perrot's introductory contribution. Historians have taken

2 While batrachia have been cloned by embryonic cell nuclear transfer since the 1950s, cloning by nuclear transfer of differentiated, "reprogrammed" somatic cells dates back to the 1990s.

up women's struggles with and against techniques. As early as 1983, Michelle Perrot focused on "machinebreakers" and "feminine luddites" (see also Jarrige (2013)). Ruth Oldenziel focused on women's historical struggles to avoid the monopoly of techniques on men. To do with or without techniques, the question was raised again in the 1970s–1980s, around life technologies and, more generally, because of the considerable extension taken by technical artifacts and equipment into the ordinary aspect of Western life, being the subject of repeated discussions opposing technoskeptics and ecofeminists, on the one hand, and rationalists and "supporters of progress", on the other hand. Today, the question of alternatives to technologies and countertechnological utopias is being revived as a sphere of interest, an object of research and a source of commitment, relegating Haraway's cyborg to a bygone era of optimism. Agnostic as to the "effects" that can be attributed to techniques, the STS movement and the feminist critique of techniques encourage us, above all, to multiply investigations and accounts of how we have developed and are developing society with techniques to open up the field to alternatives.

9.12. References

Abbat, J. (2012). *Recoding Gender. Women's Changing Participation in Computing*. MIT Press, Cambridge.

Akrich, M. (1987). Comment décrire les objets techniques ? *Techniques et culture*, 49–63.

Akrich, M. (2002). Du genre aux genres : vers la dénaturalisation d'une catégorie sociale ? In *L'engendrement des choses. Des hommes, des femmes et des techniques*, Chabaud-Rychter, D. and Gardey, D. (eds). Editions des archives contemporaines, Paris.

Akrich, M., Callon, M., Latour, B. (eds) (2006). *Sociologie de la traduction : textes fondateurs*. Mines ParisTech Les Presses, Paris.

Auslander, L. (1996). *Taste and Power: Furnishing Modern France*. University of California Press, Berkeley.

Auslander, L. (2010). *Des révolutions culturelles. La politique du quotidien en Grande-Bretagne, en Amérique et en France. XVIIe-XVIIIe siècle*. Presses universitaires du Mirail, Toulouse.

Auslander, L. (2014). Culture matérielle, histoire du genre et des sexualités. *Clio. Femmes, Genre, Histoire*, 40, 171–195.

Bajeux, C. (2022). Quelle andrologie ? Histoire des savoirs et des pratiques médicales du masculin en France et Suisse romande (années 1890-1970). PhD Thesis, Université de Genève, Geneva.

Berner, B. (1997). L'ingénieur ou le génie du mâle : masculinité et enseignement technique au tournant du XXe siècle. *Cahiers du Gedisst*, 19, 7–25.

Bijker, W. (1995). *Of Bicycles, Bakelites, and Bulbs: Toward a Theory of Sociotechnical Change*. MIT Press, Cambridge.

Bijker, W., Hughes, T.P., Pinch, T. (eds) (1987). *The Social Construction of Technological Systems*. MIT Press, Cambridge.

Bix, A.S. (2014). *Girls Coming to Tech! A History of American Engineering Education for Women*. MIT Press, Cambridge.

Braverman, H. (1974). *Labor and Monopoly Capital: The Degradation of Work in the Twentieth Century*. Monthly Review Press, New York/London.

Butler, J. (2006). *Trouble dans le genre*. La Découverte, Paris.

Canel, A. and Zachmann, K. (1997). Gaining access, crossing boundaries: Women in engineering in a comparative perspective. *History and Technology*, 14, 1–5.

Chabaud-Rychter, D. and Gardey, D. (eds) (2002). *L'engendrement des choses. Des hommes, des femmes et des techniques*. EAC, Paris.

Chanteux, A. (2009). Les inventives. Femmes, inventions et brevets en France à la fin du XIXe siècle. Documents pour l'histoire des techniques.

Cochoy, F. (1999). *Une histoire du marketing : discipliner l'économie de marché*. La Découverte, Paris.

Cockburn, C. (1991). *Brothers: Male Dominance and Technological Change*. Pluto Press, London.

Cockburn, C. (2004). Le matériel dans le pouvoir masculin. *Cahiers du Genre*, 1(30), 89–121.

Cockburn, C. and Ormrod, S. (1993). *Gender and Technology in the Making*. Sage, London.

Coffin, J. (1996). Consumption, production, and gender: The sewing machine in nineteenth-century France. In *Gender and Class in Modern Europe*, Frader, L. and Rose, S. (eds). Cornell University Press, Ithaca.

Cohen, Y. and Pestre, D. (1998). Présentation. *Annales. Histoire, Sciences Sociales*, 4(5), 721–744.

Cowan, R.S. (1983). *More Work for Mother: The Ironies of Household Technology from the Open Hearth to the Microwave*. Basic Books, New York.

Creager, A., Lunbeck, E., Schiebinger, L. (eds) (2001). *Feminism in Twentieth Century Science, Technology, and Medicine*. University of Chicago Press, Chicago.

Daune-Richard, A.M. (2003). La qualification dans la sociologie française : en quête des femmes. In *Le travail du genre. Les sciences sociales du travail à l'épreuve des différences de sexe*, Laufer, J., Marry, C., Maruani, M. (eds). La Découverte, Paris.

Dauphin, C. and Gardey, D. (2003). Introduction. In *Le travail du genre. Les sciences sociales du travail à l'épreuve des différences de sexe*, Laufer, J. (ed.). La Découverte, Paris.

Delphy, C. (1998). *L'ennemi principal (Tome 1). Économie politique du patriarcat.* Syllepse, Paris.

Edgerton, D. (1998). De l'innovation aux usages. Dix thèses éclectiques sur l'histoire des techniques. *Annales. Histoire, Sciences Sociales*, 53(4–5), 815–837.

Edgerton, D. (2007). *The Shock of the Old. Technology and Global History Since 1900.* Oxford University Press, Oxford.

Edwards, P.N. (2003). Industrial genders: Soft/hard. In *Gender and Technology*, Lehrman, N.E., Oldenziel, R., Mohun, A.P. (eds). Johns Hopkins University Press, Baltimore.

Franklin, S. and Ragoné, H. (1998). *Reproducing Reproduction: Kinship, Power and Technological Innovation*. University of Pennsylvania Press, Philadelphia.

Hackett, E., Amsterdamska, O., Lynch, M., Wajcman, J. (eds) (2008). *The Handbook of Science and Technology Studies*. MIT Press, Cambridge/London.

Harden Chenut, H. (2010). *Les Ouvrières de la République. Les bonnetières de Troyes sous la IIIe République*. PUR, Rennes.

Gardey, D. (1998). Perspectives historiques sur le travail des femmes. In *Les nouvelles frontières de l'inégalité. Hommes et femmes sur le marché du travail*, Maruani, M. (ed.). La Découverte, Paris.

Gardey, D. (2002). Humains et objets en action. Essai sur la réification de la domination masculine. In *L'engendrement des choses. Des hommes, des femmes et des techniques*, Chabaud-Rychter, D. and Delphine, G. (eds). Editions des archives contemporaines, Paris.

Gardey, D. (2008). *Ecrire, calculer, classer. Comment une révolution de papier a transformé les sociétés contemporaines (1800-1940)*. La Découverte, Paris.

Gardey, D. (2013). Le travail des femmes en France et en Grande-Bretagne de la Révolution industrielle à la Seconde Guerre Mondiale. In *Travail et genre dans le monde*, Maruani, M. (ed.). La Découverte, Paris.

Gardey, D. (2015). Gender-technology relations in the various ages of information societies. In *Connecting Women. History of Computing*, Schafer, V. and Thierry, B. (eds). Springer, Cham.

Gardey, D. (2016). "Cyborg". In *Encylopédie critique du genre*, Rennes, J. (ed.). La Découverte, Paris.

de Grazia, V. (1996). *The Sex of Things: Gender and Consumption in Historical Perspective*. University of California Press, Berkeley.

Hämmerle, C. (2018). Entretien avec Karin Hausen [Online]. Available at: http://journals.openedition.org/genrehistoire/3905 [Accessed 4 May 2022].

Haraway, D. (2007). Manifeste cyborg : science, technologie et féminisme socialiste à la fin du XXe siècle. In *Manifeste cyborg et autres essais*, Haraway, D. (ed.). Exils, Paris.

Hausen, K. (2012). *Geschlechtergeschichte als Gesellschaftsgeschichte*. Vandenhoeck & Ruprecht, Göttingen.

Hausman, B. (1995). *Changing Sex: Transsexualism, Technology and the Idea of Gender*. Duke University Press, Durham.

Hilaire-Pérez, L. (2000). *L'invention technique au siècle des Lumières*. Albin Michel, Paris.

Horowitz, R. (2001). *Boys and Their Toys? Masculinity, Class, and Technology in America*. Routledge, London/New York.

Horowitz, R. and Mohun, A. (1998). *His and Hers: Gender, Consumption and Technology*. University of Virginia Press, Charlottesville.

Jarrige, F. (2013). Le genre des bris de machines : violence et mécanisation à l'aube de l'ère industrielle (Angleterre-France, 1750-1850). *Clio. Femmes, Genre, Histoire*, 38, 17–40.

Jarrige, F. (2018). Genèse de la machine à coudre. Les Cahiers d'Adiamos, 89, 7–26.

Kergoat, D. (1982). *Les ouvrières*. Syros, Paris.

Kingery, D.W. and Lubar, S. (1995). *History from Things: Essays on Material Culture*. Smithsonian Books, Washington, D.C.

Lehrman, N.E., Mohun, A.P., Oldenziel, R. (2002). L'histoire des techniques et la question du genre. In *L'engendrement des choses. Des hommes, des femmes et des techniques*, Chabaud-Rychter, D. and Gardey, D. (eds). Editions des archives contemporaines, Paris.

Lehrman, N.E., Oldenziel, R., Mohun, A.P. (eds) (2003). *Gender and Technology: A Reader*. Johns Hopkins University Press, Baltimore.

Lerman, N.E., Mohun, A.P. and Oldenziel, R. (1997). Versatile tools: Gender analysis and the history of technology. *Technology and Culture*, 38(1) 1–8.

Lie, M. (2003). Gender and ICT – New connections. In *He, She and IT Revisited: New Perspectives on Gender and Information Society*, Lie, M. (ed.). Gylendal Akademish, Oslo.

Light, J. (1999). When computers were women. *Technology and Culture*, 40(3), 455–483.

Lipartito, K. (1991). When women were switches: Technology, work and gender in the telephone industry (1890-1930). *American Historical Review*, 99(4), 1075–1111.

MacKenzie, D. and Wajcman, J. (1985). *The Social Shaping of Technology*. Open University Press, Milton Keynes.

Marks, L. (2001). *Sexual Chemistry. A History of the Contraceptive Pill*. Yale University Press, New Haven/London.

Maruani, M. and Nicole, C. (1991). *Au labeur des dames*. Syros, Paris.

McGaw, J. (1982). Women and the history of American technology. *Signs*, 7, 798–828.

McGaw, J. (2003). Why feminine technologies matter. In *Gender and Technology. A Reader*, Lerman, N., Oldenziel, R., Mohun, A. (eds). John Hopkins University Press, Baltimore/London.

Mohun, A. (2001). Designed for thrills and safety: Gender, technology, and the commodification of risk in the amusement park industry. *Journal of Design History*, 14, 291–306.

Morley, C. and McDonnell, M. (2015). The gendering of the computing field in Finland, France and the United Kingdom Between 1960 and 1990. In *Connecting Women. History of Computing*, Schafer, V. and Thierry, B. (eds). Springer, Cham.

Noble, D. (1984). *Forces of Production: A Social History of Industrial Automation*. Knopf, New York.

Oldenziel, R. (1996). Objections. In *Learning from Things. Method and Theory of Material Culture Studies*, Kingery, D. (ed.). Smithnonian Institution Press, Washington, D.C.

Oldenziel, R. (1997). Boys and their toys: The fisher body craftsman's guild, 1930-1968, and the making of a male technical domain. *Technology and Culture*, 38(1), 60–96.

Oldenziel, E. (1999). *Making Technology Masculine: Men, Women, and Modern Machines in America, 1870-1945*. Amsterdam University Press, Amsterdam.

Oldenziel, R. and Hård, M. (2013). *Consumers, Tinkerers, Rebels: The People Who Shaped Europe*. Palgrave Macmillan, Basingstoke.

Oudshoorn, N. (1994). *Beyond the Natural Body: Archeology of Sex Hormones*. Routledge, London.

Oudshoorn, N. (2003). *The Male Pill: A Biography of a Technology in the Making*. Duke University Press, Durham.

Oudshoorn, N. and Pinch, T. (eds) (2003). *How Users Matter: The Co-Construction of Users and Technologies*. MIT Press, Cambridge.

Perrot, M. (1978). De la nourrice à l'employée. Travaux de femmes dans la France du XIXe siècle. *Le Mouvement Social*, 105.

Perrot, M. (1983). Femmes et machines au XIXe siècle. *Romantisme*, 41, 5–18.

Perrot, M. (1998). *Les Femmes ou les silences de l'Histoire*. Flammarion, Paris.

Pestre, D. (1995). Pour une histoire sociale et culturelle des sciences. Nouvelles définitions, nouveaux objets, nouvelles pratiques. *Annales. Histoire, Sciences Sociales*, 3, 487–522.

Peyrière, M. (1992). Un moteur électrique pour la machine à coudre : une innovation dans l'impasse. Le cas parisien 1867-1914. *Bulletin d'histoire de l'électricité*, 19–20, 73–86.

Pursell, C. (1993). Am I a lady or an engineer? The origins of the women's engineering society in Britain, 1918-1940. *Technology and Culture*, 34(1), 78–97.

Robert, H. (1980). Le machinisme et le travail féminin au XIXe siècle. PhD Thesis, Paris I, Paris.

Rose, N. (2006). *The Politics of Life Itself: Biomedicine, Power, and Subjectivity in the Twenty-First Century*. Princeton University Press, Princeton/Oxford.

Rossiter, M.W. (2003). L'effet Mathilda en sciences. *Les cahiers du CEDREF*, 11, 21–39.

Schafer, V. and Thierry, B. (eds) (2015). *Women, Gender and ICT in Europe*. Springer, New York.

Schiebinger, L. and Schraudner, M. (2011). Interdisciplinary approaches to achieving gendered innovations in science, medicine, and engineering. *Interdisciplinary Science Reviews*, 36(2), 154–167.

Scott, J.W. (1982). The mechanization of women's work. *Scientific American*, 247(3), 166–187.

Stanley, A. (1993). *Mothers and Daughters of Inventions: Notes for a Revised History of Technology*. Rutgers Metuchen, New Brunswick.

Strasser, S. (1995). *Satisfaction Guaranteed: The Making of the American Mass Market*. Smithsonian Institution Press, Washington, D.C.

Tabet, P. (1979). Les mains, les outils, les armes. *L'Homme*, 19(3–4), 5–61.

Tilly, L.A. and Scott, J.W. (2002). *Les femmes, le travail et la famille*. Petite Bibliothèque Payot, Paris.

Touraille, P. (2008). *Hommes grands, femmes petites : une évolution coûteuse. Les régimes de genre comme force sélective de l'évolution biologique*. Éditions de la Maison des Sciences de l'Homme, Paris.

Turkle, S. (1988). Computational reticence: Why women fear the intimate machine. In *Technology and Women's Voices: Keeping in Touch*, Kramarae, C. (ed.). Routledge, New York.

Turkle, S. (1995). *Life on the Screen: Identity in the Age of the Internet*. Simon & Schuster, New York/Toronto.

Veltz, P. (2000). *Le nouveau monde industriel*. Gallimard, Paris.

Yates, J. (1989). *Control through Communication. The Rise of System in American Management*. John Hopkins University Press, Baltimore.

Wajcman, J. (1991). *Feminism Confronts Technology*. Polity Press, Cambridge.

Wajcman, J. (2002). La construction mutuelle des techniques et du genre. In *L'état des recherches en technologie*, Chabaud-Rychter, D. and Gardey, D. (eds). Editions des archives contemporaines, Paris.

Wajcman, J. (2004). *TechnoFeminism*. Polity Press, Cambridge.

Wajcman, J. (2006). New connections: Social studies of science and technology and studies of work. *Work, Employment and Society*, 20(4), 773–786.

Wajcman, J. (2007). From women and technology to gendered technoscience. *Information, Commnication and Society*, 10(3), 287–298.

West, C. and Zimmerman, D. (1987). Doing gender. *Gender and Society*, 1(2), 125–151.

Zylberberg-Hocquart, M.H. (2002). L'aiguille, outil du féminin. In *L'engendrement des choses*, Chabaud-Rychter, D. and Gardey, D. (eds). EAC, Paris.

10

History of Techniques and Anthropology: Measure, Means, Mirror

Nathan SCHLANGER
École nationale des chartes, Paris, France

> When we claim to establish a new moral or social order,
> whether we like it or not, we are
> the representatives of the automobile, the tractor and all
> the acquisitions of our material civilization.
>
> (Leroi-Gourhan 1950, p. 42)

10.1. Introduction

Anthropology, understood as the comparative and systematic study of "others", in all their diversity and ways of being, has throughout its history shown itself to be rather reluctant and ambivalent toward techniques and technology, alternating and sometimes combining rejection, indifference and infatuation. Part of this indecision undoubtedly stems from the ambiguities surrounding the notions of "techniques" and "technology" throughout the humanities and social sciences. The various semantic fields and contexts in which these terms are used are explored in greater depth in Chapter 5. In our case, let us recall that the French research tradition (which will primarily concern us here) has long reserved the term "technology" for the discipline or field of knowledge dedicated to the study and methodical description of

History of Techniques,
coordinated by Guillaume CARNINO and Xavier GUCHET.

"techniques", which are briefly defined as "efficient traditional acts" (according to Marcel Mauss's famous formula in 1935[1]) or "materially creative activities" (as André Leroi-Gourhan noted in 1952[2]). While there is obviously no need to impose or standardize these distinctions, their relevance here is obvious. After all, it must be admitted that "technology", which for some represents an epistemological approach, can also perfectly well be understood by others as an ontological object in its own right. In this usage, which is particularly widespread throughout the English-speaking world, "technology" designates sophisticated, explicit, mechanized and universal material activities, unlike "techniques", which refer to activities that are considered simple, tacit, tool-based and localized – the techniques of angling, for example, versus the technology of the trawler. Therefore, while technology is an integral part of the industrial modernity of the West, techniques are characteristic of traditional, non-globalized, premodern societies – in other words, "others", who are the perfect subjects of anthropological study.

Let us agree, then, that the possible links between anthropology and the history of techniques can take various forms. That said, let us set aside any plans to follow the rise of techniques in time and space, from the first knapped flints of deep prehistory to the ballistic missiles of the military-industrial complex – to cite the title of a book edited by Latour and Lemonnier (1994), to which we will return – through the invention of pottery, the decline of the watermill and the advent of the prepaid smartphone. With their more or less retrospective beginnings and overly edifying outcomes, such tales of historical succession and progression are already the subject of a range of easily accessible specialist publications and illustrated albums[3].

Once we have firmly established ourselves on our historiographical ground, we need to make another important distinction between the "techniques" we are talking about here and a notion which, for all its conceptual affinities, should not be taken as

---

1 Mauss gives this definition in his 1935 essay on "Les techniques du corps" ("Body techniques"). He specifies that the technical act, unlike the religious or symbolic act, "is felt by the author *as a mechanical, physical or physicochemical act*, and that it is pursued for this purpose" (see Mauss (2023, pp. 61–63) emphasis in the original work; see also Sigaut (2002)). Mauss's main writings on techniques and technology have been translated into English as of 2006.

2 See Schlanger (2023a, p. 103).

3 For the most scholarly and researched of these publications, see the following encyclopedic works: Singer et al. (1954, 1958), Forbes (1955, 1964), Daumas (1962), Gille (1978), Bassala (1988), Pounds (1989), Pollard and Gosden (2023) and Needham (1954, 2008) for a less Eurocentric perspective.

equivalent or interchangeable: that of "material culture". Material culture – a loose and polysemous term, in fact, applied to manufactured and used things invested with cultural meanings – has a clear and complex relationship with anthropology[4]. Numerous historical and theoretical assessments of this relationship already exist, whether concerning cabinets of curiosities and their outlandish clutter, museographic displays of colonial trophies in the 19th and 20th centuries, the sociological relevance of "biographies" of more or less ordinary objects, or, more recently, the lively debates regarding the possible restitution of various masterpieces and objects steeped in history to their countries of provenance or cultures of origin[5].

The aim here is not, therefore, to take up these issues again, nor to try to decide, in a prescriptive or causal terminological approach, whether material culture is only one possible or subsidiary outcome of technological activities or whether, on the contrary, it encompasses them all. Rather, the merging between "techno-*logy*" and "material culture *studies*" will enable us to better identify some of the factors and hazards that, over the last two centuries, have led to an interest (or lack thereof) in the techniques of "others". Rather than symbols, style or materiality, notions often associated with "material culture", it is anthropology, the study of "others", that we are looking at here, and more specifically, the "technical" connotations of techniques (to evoke Febvre's "technical history of techniques"), which concern industry, infrastructures, production, effort, expertise and work. In contrast to the phenomenological and aestheticist approaches characteristic of material culture studies, we will take techniques towards the fields of political economy and sustainable development.

Why, then, should we be interested in the techniques of ethnographically known peoples? What use could there be, from a historical perspective, in the "techno-*logy*" of others? Among various conceivable and topical reasons, the following pages focus on three consecutive moments, to which rather uneven, and by no means exhaustive or even systematic, attention is given. In the first phase, throughout the 19th century, techniques were perceived as a *measure* of distinction and difference between the industrialized Western powers and their colonized subjects, whose anthropology was beginning to be studied. In a second phase, roughly beginning in the first decades of the 20th century, these markers of distance became stages on an ascending ladder that needed to be climbed: the study of the

---

4 On the notion of material culture in anthropology, archaeology and history, including its museum aspects, see, in particular, Pesez (1978), Poulot (1986), Hicks (2010), Bonnot (2014), Bourgeois et al. (2018), Gerritsen and Riello (2021) and Schlanger (2021).

5 Social and colonial aspects of the issue are, among others, addressed by Appadurai (1986), Thomas (1991), Bromberger and Chevalier (1999) and Sarr and Savoy (2018).

techniques of "others" was then seen as a *means* of facilitating their access – their accession – to modern or at least globalized modes of production and consumption. In the third and more recent phase, the techniques of the "others" are proving to be, in their particularities and common aspects, a *mirror* that amplifies and reflects our own modern condition. Here, anthropological interest in the most modest "materially creative activities" – pottery, plowing and cattle breeding, examples to which we will return later on – has significant implications for our understanding of the advanced technologies and technosciences that are our distinctive hallmark.

10.2. Techniques as a measure

In addition to numerous 19th-century historians, geographers and sociologists, not only administrators and entrepreneurs of all kinds but also anthropologists perceived techniques as markers, measures of difference and distance. During the preceding era of exploration, the Western world's encounters with "others" were relatively rare and balanced, particularly in the case of the East, which was long recognized for its written traditions and the sophistication of its material productions, silk, earthenware and gunpowder. However, as Michael Adas explains in detail in *Machines as the Measure of Men* (1989), the colonial hegemony of the 19th century led to an accentuation and overvaluation of machines and technical infrastructures, the result and driving force of the "industrial revolution", as proof of civilization and progress (Adas 1989). On the one hand, techniques are associated with the science that underpins them, which is by definition universal and rational; on the other hand, techniques are the instrument for mastering nature and guaranteeing its increasing domination by (Western) man. Europe's superiority is thus demonstrated primarily by the power of its industry and infrastructure rather than by other possible criteria, such as the validity of its modes of social organization, its political philosophy or its aesthetic productions. To this was added the presumption, be it of theological or humanist origin, of a historical destiny and a civilizing mission. As a result, the expansion and imposition of new technical regimes on colonized populations represented both an inevitable and beneficial implementation of the laws of progress. Such a connivance between material power and civilizational expansion can be seen not only in the colonial terrain itself but also in the reflections and rationalizations of the scholars of the time.

In the context of the invention of the primitive society[6], an important milestone is represented, in the field of archaeology, by the "three-age paradigm" initiated by Scandinavian prehistorians in the post-Napoleonic era and then widespread in

6 As it is called by Kuper (1988).

European museums in the second half of the 19th century[7]. The Stone Age, Bronze Age, Iron Age: Prehistoric and, by extension, primitive peoples of the earth can be distinguished according to the raw materials they use, gradually becoming more efficient and diversified over time. Another universally applicable evolutionary scheme, proposed by the American lawyer and anthropologist Lewis Henry Morgan (1818–1881), has proven particularly influential. In his 1877 book *Ancient Society*, eloquently subtitled *Researches in the Lines of Human Progress from Savagery through Barbarism to Civilization*, Morgan notes (in contrast to biblical orthodoxy or racial polygenism) that humanity began its career "at the bottom of the ladder" and then progressed through the slow accumulation of experimental knowledge in various fields. First and foremost are the means of subsistence, which "have been increased and perfected by a series of successive arts, introduced at long intervals of time, and connected more or less directly with inventions and discoveries". On this basis, human history can be divided into successive "ethnic periods": during the state of "savagery" (subdivided into lower, then middle, then upper savagery), humanity is composed of hunter–gatherers who acquired the use of fire, then the bow and arrow; then, at the state of "barbarism" (again subdivided into lower, middle and upper), pottery is added, followed by the domestication of plants and animals, before the processes of smelting and metalworking are mastered, leading to the third state, "ours", that of civilization, characterized by sophisticated social and political organization, a monotheistic religion and, above all, respect for private property[8].

With the possible exception of this last point, this combination of markers and gradation was readily endorsed by philosopher and economist Karl Marx (1818–1883), while adding to it an extra dimension. In effect, Marx emphasized the constitutive contribution of labor to the evolution of humanity and, subsequently, of the historical and dialectical links that develop between productive activities and social relations. A comment on Chapter V of his 1867 *Capital* is worth quoting here at length:

> As soon as the labour process has undergone the slightest development, it requires specially prepared instruments. Thus, we find stone implements and weapons in the oldest caves. In the earliest period of human history, domesticated animals, i.e., animals that have undergone modification by means of labour, that have been bred specially, play the chief part as instruments of labour along with stones, wood, bones and shells, which have also had work done on

7 See Rowley-Conwy (2007) and Eskildsen (2012).

8 See Morgan (1877, p. 4 and Chapter 1).

> them. The use and construction of instruments of labour, although present in germination among certain species of animals, is characteristic of the specifically human labour process, and Franklin therefore defines man as "a tool-making animal". Relics of bygone instruments of labour possess the same importance for the investigation of extinct economic formations of society as do fossil bones for the determination of extinct species of animals. It is not what is made but how, and by what instruments of labour, that distinguishes different economic epochs. Instruments of labour not only supply a standard of the degree of development which human labour has attained, but they also indicate the social relations within which men work.[9]

Once the milestones had been set and the distances established, the triumphant colonizing powers now found themselves confronted with the far more complex challenge of integrating the indigenous productive forces, which had also become consuming masses, in an operational and profitable manner. From markers of superiority and dominance, techniques also become means, true tools of civilization, mobilized in a process of economic, social and cultural enhancement.

Some thinkers saw this process as a natural evolution or historical inevitability, which it is the duty and vocation of Western civilization to support. Thus, Marx himself, in a prescient commentary on the railway network in India, postulated that "once mechanization has been introduced into the locomotion system of a country which possesses iron and coal ores, it will be impossible to prevent that country from manufacturing more of them [by itself]"[10]. Applauded as a driving force for progress and civilization, mechanization seems as desirable as it is inevitable. The more mitigated version presented by Victorian anthropologist, politician and banker John Lubbock (1834–1913) goes along the same lines: as he predicted, "given time, a Polynesian can become an Englishman".

While this aphorism has the merit of toning down prejudices of inherent racial inferiority[11], it includes an important temporal caveat. During the height of colonialism, indigenous peoples were considered to be "without history", confined

9 Marx ([1867] 1976, pp. 285–286): Marx added here a suggestive footnote whereby "The least important commodities of all for the technological comparison of different epochs of production are articles of real luxury".

10 Marx (1853), cited by Headrick (1988, p. 3), and see also the analysis of Adas (1989, p. 236).

11 As did the inclusive evolutionary schemes of L. Morgan and E.B. Tylor. On the relations between technological and racial prejudices, see Adas (1989, p. 27).

"outside time" to endure the rhythms of the seasons[12]; now, on the other hand, they are granted a seemingly incompressible timeframe in which to ascend toward global modernization, mainly through the use of techniques. However, almost everywhere, access to European techniques by indigenous populations has been halting, difficult to achieve, uncertain and destabilizing, even when well-intentioned support measures are implemented, including provisions emanating from "the anthropology of techniques" (to call it retrospectively). Alongside the destruction and erosion of traditional structures, the risk remains, as Daniel Headrick noted in *The Tentacles of Progress. Technology Transfer in the Age of Imperialism* (1988), that colonized countries will become modernized economies while remaining fundamentally underdeveloped[13].

10.3. Techniques as a means

The inclusion of these previously preindustrial populations in the global economy sphere accelerated at the turn of the 20th century, when what Headrick called "the imperialism of merchandizes" was in full swing. The ease of long-distance, relatively rapid and cheap transport has brought raw materials from the tropics to the industrial centers of the North: cotton, palm and peanut oil, forestry products, food products, etc[14]. With a certain delay, the circulation of material and goods needed for new markets emerging in sub-Saharan Africa, the Middle East and Asia was set in motion. The local populations, suppliers of raw materials, were also gradually becoming informed consumers, with their own preferences and tastes, prompting many entrepreneurs to produce locally to be closer and better adapted to the market. The example of "wax" fabrics is pertinent here: inspired by the Javanese batik, which was taken up by the Dutch and marketed in West Africa from the 19th century onward, local production of these fabrics is now threatened by low-cost Chinese imports. Whether made elsewhere or locally, these consumer goods, in turn, generate a growing need for financial resources and capital and raise the question of productive, predictable, salaried work in a standardized, globalized monetary economy[15] – over and above traditional agricultural and craft production practices,

12 Here, I paraphrase the titles of important books by Wolf (1982) and Fabian (1983), while also alluding to President Nicolas Sarkozy's infamous speech in Dakar in 2007.

13 See Headrick (1988, p. 4).

14 Without forgetting, on another level mentioned in section 10.1 concerning material culture, the circulation of items described as "primitive" or "tribal" art.

15 On labor issues in Africa, see in particular Adas (1989, p. 24), Cooper (1996) and Bellucci and Eckert (2019).

let alone the acquisition of subsistence goods through exchange and barter – or even through gift, in the Maussian sense of the term.

Marcel Mauss (1872–1950) was one of those sociologists and anthropologists who perceived the utility of studying the techniques of the "others". As he published his "Essay on the gift"[16] in 1925, Mauss joined forces with physician and linguist Paul Rivet and philosopher Lucien Lévy-Bruhl to create the Institute of Ethnology at the University of Paris. Funded in large part by the Ministry of the Colonies, the institute's aim was to professionalize ethnographic studies in France and to train future colonial administrators in the "enhancement" (*mise en valeur*) of the natural and human resources they would be responsible for in their postings. In 1930, while his students were preparing for the Dakar-Djibouti mission, Mauss himself, notorious armchair anthropologist though he was, embarked on his only visit to colonial territory: three weeks touring the Moroccan protectorate, including its recently pacified regions. The brief account he gave of his trip focused primarily on patrimonial and documentary considerations, "before it is too late". However, Mauss also recognized the importance of the *Service des arts indigènes* (as set up by the former governor of the protectorate, Maréchal Hubert Lyautey) to anticipate a form of "development anthropology", which would value the technical and artisanal activities of the local populations[17]. More details on his expectations can be found in the technology courses he taught in the early 1930s. To better resist low-cost imports from the Far East, Mauss advocated the development of indigenous products and techniques "in their line". We therefore should not give "our tools" to Moroccan craftsmen, he argued, but rather "tools that are theirs, perfected". Lastly:

> we are witnessing the contact between civilizations, which is being brutally brought about by techniques. We need to find out how indigenous techniques survive, persist, disappear or are compromised. We have seen Touggala potters imitate European hardware. These studies are not of historical interest, but of current interest, providing material for the study of societies undergoing transformation.[18]

André Leroi-Gourhan (1911–1986) was also interested in the fate of these "societies in the process of transformation", even if he did not fully subscribe to the

16 See Mauss (1924).

17 See Mauss (1930).

18 See a quick overview of this journey and its implications in Schlanger (2012, pp. 118–121). Mauss's writings on techniques have been translated into English in Mauss (2006) and analyzed in Vatin (2004) and Schlanger (2012).

teachings and outlook of his mentor Mauss[19]. Leroi-Gourhan's first defamiliarizing contact was actually quite atypical, when, between 1937 and 1939, a Japanese government scholarship enabled him to spend 2 years in the archipelago. The wide range of ethnographic and technological observations he gathered focused mainly, but not exclusively, on traditional activities. Indeed, the "opening" of the Meiji era to foreign influences from the 1860s onward represented, for him, an unprecedented historical and sociological experience; that of a deliberate, managed and largely successful transition to industrial modernization (including the mass production of the junk that would flood North African markets). In fact, Leroi-Gourhan at one point considered devoting his secondary thesis to "the evolution of Japanese craft tools since 1868"[20]. In any case, the main thesis he defended in Sorbonne in June 1944, entitled *Archéologie du Pacifique-Nord. Matériaux pour l'étude des relations entre les peuples riverains d'Asie et d'Amérique* (Archaeology of the North Pacific. Materials for the study of relations between the riparian peoples of Asia and America), equally influenced by these Japanese experiences, proves revealing in both form and content.

From the point of view of its theoretical trends, this thesis, published in 1946, is set in a broader historical and ideological context, also perceptible in Leroi-Gourhan's better-known publications *L'Homme et la matière* (1943) and *Milieu et techniques* (1945)[21]. The colonial reconfiguration referred to in section 10.2, shifting from the imperatives of territorial and political domination to the administration of commercial and cultural interconnections, was not without implications for the fate of the "evolutionist" and "diffusionist" paradigms within the human and social sciences in the first half of the 20th century. Less doctrinal than other anthropological or culture-historical trends of the time – think of the English hyperdiffusionist schools, or the promoters of *Kulturkreiselehre* in the German-speaking world – Leroi-Gourhan's challenge was to know whether the circulations documented across the world (in this case, around the Pacific Ocean, from the Philippines to Peru) involved raw materials, manufactured objects, or their more or less faithful local imitation – unless these were the populations themselves that were on the move from their original homes, driven or, on the contrary, attracted to new

19 I partly rely here on the analyses presented in Schlanger (2023a), as well as on the Leroi-Gourhan archives, held at the Maison des sciences de l'Homme-Mondes, in Nanterre. On the life and work of Leroi-Gourhan, see, in particular, Soulier (2018).

20 These Japanese experiences are included in the posthumous collection *Pages oubliées sur le Japon* (Leroi-Gourhan 2004). For the "spectre of industrialized Asia" and the threats posed by industrious Japan, see Adas (1989, p. 357).

21 See Leroi-Gourhan (1943, 1945, 1946, 2025) for presentation and partial translation into English) as well as Schlanger (2023a, 2025).

living spaces? As for the similarities observed between the material practices and productions among two distinct groups, they can be understood "either as survivals of an archaic cultural fund, or as the result of simultaneous or successive exchanges, [or even as] ensembles transported by migrations". From this perspective, the study of manufacturing techniques, particularly for everyday objects, takes on a well-defined epistemological function – that of the "critique" of sources and documents, which enables us to unmask "false evidence, [that is] objects whose universality does not mean anything", and to identify those that are, on the contrary, specific, reliable and revealing[22]. These "technical convergences" are determined by the means of action on the material and the objectives to be achieved. They correspond to those inevitable rectilinear "tendencies" which – as Leroi-Gourhan puts it in *L'Homme et la matière* – always and everywhere drive "the handheld flint [tool] to acquire a handle, or the bundle dragged on two poles to equip itself with wheels"[23].

To be sure, these speculations pertain in several respects to the prevailing anthropological theorizing of the time – still essentially based on travel accounts, collections and museographic comparisons – fueled by hypotheses about the determinism of natural forces and the more or less asserted conservatism of the human species. In addition, however, these questions are clearly attuned to the zeitgeist, particularly in their colonial declensions: like other scientific disciplines concerned with the mastery and *mise en valeur* of the tropics, anthropology also focuses on the challenges of material circulations and technical transfers that characterize the march of industrial capitalism across civilizations.

This attention is obviously also reflected in the administrative and institutional sphere. In September 1944, his thesis defended, Leroi-Gourhan was appointed to the chair of *Ethnologie coloniale* (Colonial Ethnology) established at the University of Lyon on the initiative of Marcel Griaule. This decade in Lyon (until his return to Paris in 1956) was particularly productive for him, consolidating his expertise and reputation in the fields of prehistory, paleontology and cave art. However, Leroi-Gourhan did not neglect the colonial aspect of his chair, whose relevance was reinforced in 1946 by the creation of the French Union. The two volumes entitled *Ethnologie de l'Union française (territoires extérieurs)*[24] published in 1953 with Jean Poirier confirmed this interest, as did the establishment in 1946 of the *Centre de formation aux recherches ethnologiques* (CFRE)[25], in association with the *Musée*

22 See Leroi-Gourhan (1946, pp. 4–8, 510), and Leroi-Gourhan (1949, pp. 153–154).

23 See Leroi-Gourhan (1943, p. 28, translated in 2025, p. 102).

24 Ethnology of the French Union (outer territories) (Leroi-Gourhan and Poirier 1953).

25 Training Center for Ethnological Research.

de l'Homme and the CNRS[26]. While the CFRE's activities and investigations remained essentially metropolitan in scope – due to a lack of resources, but also, as we shall see, as a methodological choice – its ambition was to train ethnologists and technologists capable of managing the interface between, on the one hand, the material and economic arrangements resulting from colonization and, on the other hand, the aptitudes and appetites of natives "in the process of transformation".

The proximity of this administrative and operational context with the theoretical developments of Leroi-Gourhan's technology brings us closer to an "applied anthropology of techniques", or even "applied technology". All the more so given that the concepts of "tendency", "degree of fact" and "technical milieu", already established in 1943, were enriched and reoriented from the early 1950s onward with those of the "*chaîne opératoire*" or operational chain – a powerful concept that focuses on the various processes of transformations that lead from raw materials to finished and used products[27]. In addition to its pragmatic intentions in the socioeconomic field, this technology also has an undeniable vocation for improvement that, in our author's case, seems to combine paternalistic humanism, in the tradition of Mauss and the Institute of Ethnology, with the reaffirmation of his Catholic faith. Of the various writings and lecture notes that attest to this "applied technology", I will rely here on a little-known but particularly revealing text entitled "*Civilisation matérielle et vie spirituelle*" ("Material civilization and spiritual life"), taken from a lecture given in Lyon in July 1950 and published the same year in the missionary magazine *Rythmes du monde*.

While delineating the respective spheres of action of men of science and men of cloth, Leroi-Gourhan aims to bring material civilization as close as possible to its cultural and spiritual dimensions. He mentions that, currently, ethnology is no longer confined to the study of rare or curious peoples, but is moving toward facts "more directly linked to the development of civilizations", making it both practical and socially useful. This has also been recognized in the United States and, recently, in France: "For some years now – especially since 1946 – the French administration has turned to ethnologists to ask them for preparatory studies regarding the social

26 See Gutwirth (2010) on the history of the CFRE, and Soulier (2018, p. 257) and Schlanger (2023a, p. 228). At the same time, Leroi-Gourhan set up the *Centre de documentation et de recherches pré-historiques* (Centre for Prehistoric Documentation and Research) (CDRP).

27 On the operational chain, see, in particular, Cresswell (1983, 1996) and Balfet (1990) and references in Schlanger (2023a, pp. 381–393). André Leroi-Gourhan's main "technological" texts are collected in an anthology in English, see Leroi-Gourhan (2025) and Schlanger (2025).

and material situation of such and such group within the *Union française*"[28]. It is in direct relation to this concern for social utility, Leroi-Gourhan continues, that "technology" is emerging today, defined here as "This discipline [part] of ethnology [that] has as its goal the study and description of human technical behavior [*l'étude et la description du comportement technique des hommes*]"[29].

Technology involves material descriptions that are as precise and detailed as possible, preferably in the field (tool-making, pottery, basketry, food, etc.). However, this is only part of the story: once the material forms of human activity have been studied and classified, "technology quickly expands in the social and the economic directions. In reality, the study of any civilization leads us to establish indissoluble links between techniques, social structures, and the economy"[30]. This extension is all the more relevant when different civilizations come into contact. "When the administration claims to impose our law, or our plows, on indigenous minorities, and if it proceeds to do so without sufficient prior familiarity with the indigenous substrate, deculturation will ensue", in other words, incoherence and "social and technical maladjustments". This "technical awkwardness" afflicts certain parts of the Black world, where for example, the deep plowing thoughtlessly introduced to Fula peanut farmers in Senegal soon proved to be harmful both materially, by eroding arable land, and socially, being beyond financial reach[31]. Similarly, in Japan, and accepting that the process here was largely self-inflicted, Leroi-Gourhan was personally able to observe the social dislocations produced by the industrialization of certain districts and the creation of a working-class proletariat, owing to "changes in family structure, the very profound economic and social modifications produced by the construction of factories" – in other words, a process of "detribalization". Against such failures, Leroi-Gourhan identifies a new role for the technologist. The seasoned, positivist specialist in "human technical behavior" must also take on the task of the "interpreter":

> It now appears that one of the ethnologist's roles is to be an agent of acculturation, an agent of transmission, between different states of

---

28 (Leroi-Gourhan 1950, pp. 38–39, translated in 2025, p. 163).

29 (Leroi-Gourhan 1950, p. 39, note 1, translated in 2025, p. 163). On this important notion of "technical behavior", see Schlanger (2023a, p. 284).

30 (Leroi-Gourhan 1950, p. 40, translated in 2025, p. 164).

31 (Leroi-Gourhan 1950, p. 42 and 44, translated in 2025, p. 166 and 167). Other classic examples of deculturation induced by the hasty introduction of ill-adapted techniques include steel axes among the Yir Yoront of Australia (Sharp 1952), and see also Godelier and Garanger's (1973) study of stone and steel tools in New Guinea.

civilization. From the moment we are compelled to admit that progress in techniques is inevitable, we have to take stock of this and envisage a technology able to ensure the transitions between the [different] technical states"[32].

10.4. Techniques as mirrors

This transition between technical states unfolds as much in time as in space[33]. In the long term, technical progress seems inescapable and is "one of the paths of general evolution" that every human group is destined to follow. Echoing Lubbock's commentary on Polynesians, Leroi-Gourhan was in no doubt that "any given group, when the occasion arises, will inevitably clear the stages of material progress"[34]. Beyond this evolutionist, as well as anti-racist expectation – troubled, however, by these problematic notions of "opportunity" (which?) being "given" (by whom? how?) – Leroi-Gourhan remains more nuanced in regard to the space of techniques, or, better still, to use his own expression, the "milieu technique" ("technical milieu"). While we might expect to observe at a given location and moment in time both the homogeneity of a technical level and a continuous gradation from center to periphery, it is in fact a heterogeneous embedding that reveals itself to the anthropologist:

> In reality, the successive technical stages that range from the primitive to the most perfected are, in each civilization, nested [*s'emboîtent*] into each other. If we take our own example, we realize as we travel the French countryside that, in the same village, five or six stages of technical development can often be found superimposed. Between the hoe, still in use in some agricultural practices, and the tractor there is a distance that, from a material point of view, is the same as the one that separates the Fula people of Sudan from ourselves[35].

---

32 (Leroi-Gourhan 1950, pp. 43–45, translated in 2025, p. 167).

33 It should be noted, however, that this technology "capable of ensuring transitions" does not really raise any questions about its possible collusion with the forces at work. Leroi-Gourhan's desire for objective impartiality, his acceptance of the inevitability of progress, his manifest "technicity", in his eyes, all appear to clear him of any risk of scientific and ethical compromise, even though the aim is to establish social, economic and indeed technical control over "others".

34 (Leroi-Gourhan 1950, p. 39, translated in 2025, p. 163).

35 (Leroi-Gourhan 1950, p. 39, translated in 2025, pp. 163–164). He also mentions the Breton village, where some farmers use a hand-operated grindstone to process buckwheat, while others have reached motorization.

The conclusion Leroi-Gourhan drew from this proximity is essentially methodological, insofar as "technology cannot be understood as a strict and narrow study of the techniques of material life". Undoubtedly. For our part, however, let us note here a clear statement that places the apparently tangled and asynchronous technical interlockings of the African peasant and the French countryman on the same plane, making their equivalence or common scale quite tangible – a "symmetry", as some might call it, between "us" and "the others". What has long been the stuff of humanist aspirations or well-intentioned reflections is now within reach, owing to the study of "materially creative activities". It may well be that this harmonization is in part strategic and institutional, insofar as the investigations carried out in the corners of Morbihan (under the aegis of CFRE) could be deemed to be perfectly adequate and formative training for missions to be undertaken in Fouta-Djallon, within the more general framework of "development technology" across the French Union. However, it remains that this equivalence is also anthropological in the broadest sense of the term, insofar as it undermines and challenges previously upheld dualisms and "grand divides" – and notably does so through the study of techniques, which were until recently hailed as the distinctive preserve of Western civilization.

"Technological problems", noted by Leroi-Gourhan, "are of immediate relevance not only when we are talking about far-flung countries but also when we are dealing with our own terrain, our own soil"[36]. This immediacy, although rich in mediation, is illustrated (again in his 1950 article) through a striking example: the study launched by CFRE in 1949 of cattle breeding in a village in Ille-et-Vilaine on the border between the distribution zones of Norman and Breton cows. As this seemingly "technical", taxonomic and cartographic investigation proceeds, it turns out that those are the Breton cows, recently introduced by a new landlord, that are more prosperous and healthier than the Normandes, which have a better reputation. This outlined "bovine sociology" (sic) may therefore be seen as the beginnings of a second, stronger "symmetry", which would place humans and non-humans on the same pasture, so to speak, and which would relate the cementing of slurry pits and the ensuing heightened hygiene to the Breton tastes of the local land owner. Enjoying the meandering, offbeat nature of his own observations, the technologist Leroi-Gourhan readily noted that:

> This investigation, launched from the strict perspective of the study of material culture [*stricte étude de la culture matérielle*], has finally led me, as a last development of the questions regarding the cow

36 (Leroi-Gourhan 1950, p. 41, translated in 2025, p. 165).

> distribution map, to an in-depth study regarding attendance to either free [i.e., religious] or secular schools in the local area[37].

The anthropological study of techniques thus reveals a "closely knit network of political, social, economic and religious considerations". In this dense and heterogeneous entanglement, we can grasp the two "symmetries" indicated above, first, that between ourselves and others and second, that between humans and non-humans. Of course, there is no question here of producing facile and often fake filiations or of recruiting unwitting or even reluctant "precursors". Nevertheless, in conclusion, let us agree that the anthropological study of techniques among the "others" has become, following the research tradition presented here, an illuminating and magnifying mirror for the study of "our" modern technologies and technosciences in the wake of recent social studies of science and other science and technology studies (STSs).

The "*chaîne opératoire*" (operational chain) here represents a useful link, owing primarily to the work of Robert Cresswell (1922–2016). A close student of Leroi-Gourhan already at Lyon, a graduate and then supervisor at the CFRE, Cresswell took over ethnology teaching at Sorbonne and established the research team and journal *Techniques and Culture* in the 1970s. The coupling of "Transfer of techniques and operational chains" highlighted by Cresswell across several studies seems to be a direct extension of this technological approach to "milieus" and "transitions"[38]. For Cresswell, technologists cannot be satisfied with banalities when they ask practical questions about development or transfer techniques from high-tech countries to countries with traditional techniques. Instead, the technologist needs "a working tool – the *chaîne opératoire*, then – that both links technical and social relations and enables the internal study of a technical process". More broadly, the anthropologist must "first, demand the study of systems and structures, and not simply that of objects and tools, and second, emphasize that technical relations are inevitably accompanied by social relations within these same structures"[39]. From Irish potato farmers to Lebanese olive presses, the "cultural technology" developed by Cresswell[40] saw itself amplified by Pierre Lemonnier's "anthropology of

37 (Leroi-Gourhan 1950, p. 41, translated in 2025, p. 165)

38 Although Leroi-Gourhan himself, for various reasons of substance and circumstance, had rather turned away from the anthropology of techniques and reoriented himself, even before the publication of *Le Geste et la parole* in 1964 and 1965 and his election to the Collège de France in 1969, toward prehistoric excavation methods and the study of Paleolithic art. See Leroi-Gourhan (1964, 1965).

39 (Cresswell 1983, pp. 143, 144 and 146).

40 (Cresswell 1983, 1996) and see the work of the *Techniques and Culture* team (Coupaye and Douny 2009; Bartholeyns et al. 2010).

technical systems". Indeed, Lemonnier's studies of Guérande salt production led him to the gardening and enclosure practices of the Anga people of Papua New Guinea. By way of return, he could show that the collective representations held by these highlanders of material efficiency, as incorporated, for example, into the design of eel traps, could resonate with "our" own collective representations when we accept – or reject – the supposedly over-determined designs of jet aircrafts[41].

These mirror effects, with their accompanying changes in focus, amplifications and dynamics, converge on the study of Western technologies. Undoubtedly, as Mauss in Morocco and Leroi-Gourhan in Japan have experienced, appreciating the techniques of the "others" is never indifferent or unrelated to "our" Western modernity. Recently, an even closer connection has been made with STSs[42]. Drawing on his anthropological journeys between the Breton marshes and the highlands of Papua, Lemonnier can readily point out in his influential edited volume *Technological Choices. Transformation in Material Cultures Since the Neolithic* (1993) that "the dependence of technical design and processes upon, and their embedding in, non-technical and wider social logics is definitely not peculiar to primitive or non-industrial societies"[43]. On the contrary, such a constitutive embedding of the non-technical in the technical can also be found, upon closer inspection, in industrialized Western societies. This observation is not lost to sociologist Madeleine Akrich, who discusses in the same volume the tribulations of a technosocial transfer in Costa Rica. She notes that "the case of developing countries, which are characterized by a maximal distance between the context in which the technology supply is developed (since technologies are in general conceived in industrialized countries) and the context of the demand, permits us to see the limitations of [the] two approaches" which she criticizes, namely, "technologism" and "sociologism"[44]. Akrich is thus looking for a "distance" effect, a physical and conceptual space where the uses of techniques have not yet entered the mainstream, with forces and potential actors still to be harnessed. In this way, the gaze of the observer from outside a given culture – the ethnologist's gaze, in other words – makes it possible to distinguish between technical objects and their uses and thus proceed with a genuine technological analysis[45].

---

41 See Lemonnier (1986, 1992, 2012).

42 On the subject of technical transfer, we could also mention the "anthropotechnological" work of Wisner (1979, 1997) at an ergonomic level, and at the more sociological level, that of Geslin (1999, 2017), as well as the anthropology of "sociotechnical systems" developed by Pfaffenberger (1988, 1992).

43 (Lemonnier 1993, p. 6).

44 (Akrich 1993, p. 289).

45 (Akrich 1993, pp. 290–291).

In the same vein, further in this volume, the "ethnographer of machines" that was philosopher Bruno Latour (1947–2022) makes a plea for a "symmetrical anthropology of techniques", whose starting point is precisely these societies that truly have never been modern, nor ever wished to be. "By relocating the work of producing truth, efficiency and profitability, it is not only the premodern societies but also *our own world* that takes on a new aspect. To begin with, our own world stops being modern because it no longer differs radically from the others"[46]. Far from being torn apart by an overwhelming dualistic "great divide", this world is instead tugged across its seams by a multitude of small, more or less ad hoc or permanent divides, "assemblages" that actually become intelligible through their rigorous observation and empirical study, such as the intriguing "bovine sociology" mentioned in section 10.4.

Although this is not the only conceivable path or one effectively taken, it appears then that it is through anthropology – that is, the comparative and systematic study of "others" – that the *social intelligence of techniques* can now be understood. This is, in any case, the subtitle that Latour and Lemonnier gave to their book, *De la préhistoire aux missiles balistiques* (*From Prehistory to Ballistic Missiles*), the result of a seminar held at the Fondation des Treilles in 1992[47]. More generally, with the experience of otherness and defamiliarization that it has brought and with the challenges of transfer and intelligibility that it has had to address (with admittedly uneven success), the anthropology of techniques has proved to be beneficial on several counts and promises to continue to be so. Anthropology has made it possible to untangle techniques from their long too-obvious and overwhelming associations with the objective forces of nature and the determinisms of progress. Anthropology encourages us to understand techniques as social and cultural phenomena, among others, and even more so as composites or assemblages, as artifacts (so to speak), to be understood historically, sociologically and, of course, anthropologically.

10.5. References

Adas, M. (1989). *Machines as the Measure of Men. Science, Technology and Ideologies of Western Dominance*. Cornell University Press, Ithaca.

Akrich, M. (1993). A gasogene in Costa Rica. An experiment in techno-sociology. In *Technological Choices. Transformations in Material Cultures Since the Neolithic*, Lemonnier, P. (ed.). Routledge, London.

46 (Latour 1993, p. 378).

47 (Latour and Lemonnier 1994).

Appadurai, A. (ed.) (1986). *The Social life of Things*. Cambridge University Press, Cambridge.

Balfet, H. (ed.) (1991). *Observer l'action technique. Des chaînes opératoires, pour quoi faire ?* CNRS Éditions, Paris.

Bartholeyns, G., Govoroff, N., Joulian, F. (eds) (2010). Cultures matérielles. *Techniques and Culture*, 54–55.

Bassala, G. (1988). *The Evolution of Technology*. Cambridge University Press, Cambridge.

Bellucci, S. and Eckert, A. (2019). The "Labour Question" in Africanist historiography. In *General Labour History of Africa. Workers, Employers and Governments, 20th–21st Centuries*, Bellucci, S., Eckert, A. (eds). James Currey, London.

Bonnot, T. (2014). *L'attachement aux choses*. CNRS Éditions, Paris.

Bourgeois, L., Alexandre-Bidon, D., Feller, L., Mane, P., Verna, C., Wilmart, M. (eds) (2018). La culture matérielle : un objet en question : anthropologie, archéologie et histoire. In *Actes du colloque international de Caen (9 et 10 octobre 2015)*. Presses universitaires de Caen, Caen.

Bromberger, C. and Chevalier, D. (eds) (1999). *Carrières d'objets, innovations et relances*. Éditions de la MSH, Paris.

Cooper, F. (1996). *Decolonization and African Society. The Labor Question in French and British Africa*. Cambridge University Press, Cambridge.

Coupaye, L. and Douny, L. (2009). Dans la trajectoire des choses. Comparaison des approches francophones et anglophones contemporaines en anthropologie des techniques. *Techniques and Culture*, 52–53, 12–39.

Cresswell, R. (1983). Transferts de technique et chaînes opératoires. *Techniques and Culture*, 2, 143–163.

Cresswell, R. (1996). *Prométhée ou Pandore ? Propos de technologie culturelle*. Kimé, Paris.

Daumas, M. (ed.) (1962). *Histoire générale des techniques, vol. 1 : les origines de la civilisation technique*. Presses Universitaires de France, Paris.

Eskildsen, K. (2012). The language of objects: Christian Jürgensen Thomsen's science of the past. *Isis*, 103(1), 24–53.

Fabian, J. (1983). *Time and the Other: How Anthropology Makes Its Object*. Columbia University Press, New York.

Forbes, R.J. (1955–1964). *Studies in Ancient Technology*. Brill, Leiden.

Gerritsen, A. and Riello, G. (eds) (2021). *Writing Material Culture History*. Bloomsbury Academic, London.

Geslin, P. (1999). *L'apprentissage des mondes, une anthropologie appliquée aux transferts de technologies*. Éditions de la Maison des Sciences de l'Homme, Paris.

Geslin, P. (ed.) (2017). *Inside Anthropotechnology: User and Culture Centered Experience.* ISTE Ltd, London, and John Wiley and Sons, New York.

Gille, B. (ed.) (1978). *Histoire des techniques*. Gallimard, Paris.

Godelier, M. and Garanger, J. (1973). Outils de pierre, outils d'acier chez les Baruya de Nouvelle-Guinée. *L'Homme*, 13(3), 187–220.

Gutwirth, J. (2001). La professionnalisation d'une discipline : le centre de formation aux recherches ethnologiques. *Gradhiva*, 21, 24–41.

Headrick, D.R. (1988). *The Tentacles of Progress. Technology Transfer in the Age of Imperialism, 1850-1940*. Oxford University Press, Oxford.

Hicks, D. (2010). The material culture turn. Event and effect. In *The Oxford Handbook of Material Culture Studies*, Hicks, D. and Beaudry, M. (eds). Oxford University Press, Oxford.

Kuper, A. (1988). *The Invention of Primitive Society. Transformations of an Illusion.* Routledge, London.

Latour, B. (1993). Ethnography of a "high-tech" case. About Aramis. In *Technological Choices. Transformations in Material Cultures Since the Neolithic,* Lemonnier, P. (ed.). Routledge, London.

Latour, B. and Lemonnier, P. (eds) (1994). *De la préhistoire aux missiles balistiques. L'intelligence sociale des techniques*. La Découverte, Paris.

Lemonnier, P. (1986). The study of material culture today: Towards an anthropology of technical systems. *Journal of Anthropolgical Archaeology*, 5, 147–186.

Lemonnier, P. (1992). *Elements for an Anthropology of Technology*. University of Michigan Museum of Anthropology, Ann Arbor.

Lemonnier, P. (1993). Introduction. In *Technological Choices. Transformations in Material Cultures Since the Neolithic*, Lemonnier, P. (ed.). Routledge, London.

Lemonnier, P. (2012). Des objets pour penser l'indicible. La nécessaire convergence des théories de la culture matérielle. In *La Préhistoire des autres. Perspectives archéologiques et anthropologiques*, Schlanger, N. and Taylor, A.-C. (eds). La Découverte, Paris.

Leroi-Gourhan, A. (1943). *Évolution et techniques 1 : l'Homme et la matière*. Albin Michel, Paris. Partly translated in 2025, 85–130 [Online] Available at: https://haubooks.org/leroi-gourhan_technology/.

Leroi-Gourhan, A. (1945). *Évolution et techniques 2 : milieu et techniques*. Albin Michel, Paris. Partly translated in 2025, 131–149 [Online]. Available at: https://haubooks.org/leroi-gourhan_technology/.

Leroi-Gourhan, A. (1946). *Archéologie du Pacifique-Nord. Matériaux pour l'étude des relations entre les peuples riverains d'Asie et d'Amérique*. Travaux et mémoires de l'Institut d'ethnologie, Paris.

Leroi-Gourhan, A. (1949). Note sur les rapports de la technologie et de la sociologie. *L'Année sociologique*, 2, 766–772. Translated in 2025, 151–158 [Online]. Available at: https://haubooks.org/leroi-gourhan_technology/.

Leroi-Gourhan, A. (1950). Civilisation matérielle et vie spirituelle. *Rythmes du monde*, 3, 38–45. Translated in 2025, 159–170 [Online]. Available at https://haubooks.org/leroi-gourhan_technology/.

Leroi-Gourhan, A. (1964). *Le geste et la parole 1 : technique et langage*. Albin Michel, Paris. Translated in 1993 as *Gesture and Speech*. Translation by Anna Bostock Berger and Introduction by Randall White. October Books/MIT Press, Cambridge, MA.

Leroi-Gourhan, A. (1965). *Le geste et la parole 2 : la mémoire et les rythmes*. Albin Michel, Paris. Translated in 1993 as *Gesture and Speech*. Translation by Anna Bostock Berger and Introduction by Randall White. October Books/MIT Press, Cambridge, MA.

Leroi-Gourhan, A. (2004). *Pages oubliées sur le Japon, recueil posthume établi et présenté par Jean-François Lesbre*. Jérôme Millon, Grenoble.

Leroi-Gourhan, A. (2025). *André Leroi-Gourhan on Technology: A Selection of Writings from the 1930s to the 1960s* [Online]. Available at: https://haubooks.org/wp-content/uploads/2025/03/Andre%CC%81-Leroi-Gourhan-on-Technology_2025_Bard_HAU-Books.pdf..

Leroi-Gourhan, A. and Poirier, J. (1953). *Ethnologie de l'Union française. Volume 1: Afrique. Volume 2: Asie, Océanie, Amérique*. Presses Universitaires de France, Paris.

Marx, K. (1976). *Capital. A Critique of Political Economy. Volume 1*. Penguin Books/New Left Review, Hammondsworth [Original published 1867].

Mauss, M. (1924). Essai sur le don. Forme et raison de l'échange dans les sociétés archaïques. *L'Année sociologique*, 1, 30–186.

Mauss, M. (1930). Exposé sur un voyage au Maroc. Compte rendu à l'Institut français d'anthropologie. *L'Anthropologie*, 40, 453-456.

Mauss, M. (2006). *Techniques, Technology and Civilisation*. Durkheim Press/Berghahn Books, Oxford.

Mauss, M. (2012). *Techniques, technologie et civilisation*. Presses Universitaires de France, Paris.

Mauss, M. (2023). *Les techniques du corps*. Flammarion, Paris.

Morgan, L.H. (1877). *Ancient Society, or Researches in the Lines of Human Progress from Savagery through Barbarism to Civilization*. Macmillan, London.

Needham, J. (ed.) (1954–2008). *Science and Civilisation in China*. Cambridge University Press, Cambridge.

Pesez, J.-M. (1978). Histoire de la culture matérielle. In *La nouvelle histoire*, Le Goff, J. (ed.). Retz, Paris.

Pfaffenberger, B. (1988). Fetishised objects and humanised nature: Towards an anthropology of technology. *Man*, 23(2), 236–252.

Pfaffenberger, B. (1992). Social anthropology of technology. *Annual Review of Anthropology*, 21, 491–516.

Pollard, A.M. and Gosden, C. (2023). *An Archaeological Perspective on the History of Technology*. Cambridge University Press, Cambridge.

Poulot, D. (1997). Une nouvelle histoire de la culture matérielle ? *Revue d'histoire moderne et contemporaine*, 44(2), 344–357.

Pounds, N. (1989). *Hearth and Home: A History of Material Culture*. Indiana University Press, Bloomington.

Rowley-Conwy, P. (2007). *From Genesis to Prehistory. The Archaeological Three Age System and Its Contested Reception in Denmark, Britain and Ireland*. Oxford University Press, Oxford.

Sarr, F. and Savoye, B. (2018). *Rapport sur la restitution du patrimoine culturel africain. Vers une nouvelle éthique relationnelle*. Le Seuil, Paris.

Schlanger, N. (2012). Une technologie engagée : Marcel Mauss et l'étude des techniques dans les sciences sociales. In *Techniques, technologie et civilisation*, Mauss, M. (ed.). Presses Universitaires de France, Paris.

Schlanger, N. (2021). Matter(s) of material culture (Afterword). In *The Material Subject. Rethinking Bodies and Objects in Motion*, Mohan, U. and Douny, L. (eds). Routledge, London.

Schlanger, N. (2022). La technologie de Leroi-Gourhan : les enchevêtrements d'un chef de file. *Artefact*, 15, 259–279.

Schlanger, N. (2023a). *L'invention de la technologie. Une histoire intellectuelle avec André Leroi-Gourhan*. Presses Universitaires de France, Paris.

Schlanger, N. (2023b). Archaeology is but ethnology in the past tense. Theoretical proofs and intellectual technologies in André Leroi-Gourhan's archived *Archéologie du Pacifique-Nord*, 1946. *Bulletin of the History of Archaeology*, 33(1), 1–14.

Schlanger, N. (2025). Reading technology. Introducing Leroi-Gourhan. In *André Leroi-Gourhan on Technology: A Selection of Writings from the 1930s to the 1960s*. Bard Graduate Center, New York/Hau Books, The University of Chicago Press, Chicago.

Sharp, L. (1952). Steel axes for stone-age Australians. *Human Organization*, 11(2), 17–22.

Sigaut, F. (2002). La formule de Mauss. *Techniques and Culture*, 40, 153–168.

Singer, C. (ed.) (1954–1958). *A History of Technology*. Clarendon Press, Oxford.

Soulier, P. (2018). *André Leroi-Gourhan (1911-1986), une vie*. CNRS Editions, Paris.

Thomas, N. (1991). *Entangled Objects: Exchange, Material Culture, and Colonialism in the Pacific*. Harvard University Press, Harvard.

Vatin, F. (2004). Mauss et la technologie. *Revue du MAUSS*, 23, 418–433.

Wisner, A. (1979). *Vers une anthropotechnologie : comment pourvoir les pays en développement industriel de machines et d'usines qui marchent*. Conservatoire National des Arts et Métiers, Paris.

Wisner, A. (1997). *Anthropotechnologie : vers un monde industriel pluricentrique*. Octares Éditions, Toulouse.

Wolf, E. (1982). *Europe and the People without History*. University of California Press, Berkeley/Los Angeles.

11

Philosophy and History of Techniques

Xavier Guchet
COSTECH, Université de technologie, Compiègne, France

11.1. Introduction

When did the philosophy of technology begin? Opinions on this matter differ. If we look at the institutional aspects, we have to return to the mid-1970s, with the creation of a learned society known as the Society for Philosophy and Technology (SPT), which quickly started a journal called *Research in Philosophy and Technology*, whose first issue appeared in 1978[1]. The discipline's other major journal, *Philosophy and Technology*, was launched in 1987. A philosophy of technology department was set up in a number of universities, especially in northern Europe and North America. Thus, for almost 50 years, the philosophy of technology has been recognized as a discipline in its own right, independent of its big sister, the philosophy of science.

Notably, it was not until the 1970s that philosophers became interested in technique. Plato had already developed an elaborate concept of the arts (*technai*). Subsequently, Aristotle, Descartes, Leibniz, Diderot, Comte, Marx, Bergson and many others considered technique to be an important philosophical issue. However, it was only in the 20th century that technique came to be understood as a global phenomenon, affecting the human condition in all its aspects, justifying the birth of

1 After ceasing publication between 1985 and 1988, the journal was relaunched and became the digital journal *Techné: Research in Philosophy and Technology* in 1995.

History of Techniques,
coordinated by Guillaume Carnino and Xavier Guchet.

a philosophy of technology, as it was in the first half of the 20th century. According to Carl Mitcham, its founders are Heidegger, Jaspers, Ortega y Gasset and Dewey (Mitcham 1994). In short, there is disagreement as to when the philosophy of technology was born. This is due to the opposition between two conceptions of technique as a philosophical problem: on the one hand, it is a special problem, the object of a specialized discipline, which leads us to locate this discipline's inception precisely at the moment when it became institutionalized, in the 1970s; on the other hand, it is an introduction to questions of general philosophy, which leads us, on the other hand, to locate its beginnings earlier, with philosophers who certainly attached great importance to technique but who were certainly not "philosophers of technology" stricto sensu, in other words, in the sense of representatives of a specialized discipline (which neither Heidegger nor Jaspers nor Ortega nor Dewey were).

Thus, the birth of the philosophy of technology varies according to one's conception of technique as the object of a specialized discipline or as a question of general philosophy. Does this duality of perspectives also translate into competing conceptions of the relationship between the philosophy of technology and the human and social sciences – primarily the history of techniques? This is a reasonable assumption, given the respective positions of these two major conceptions of the philosophy of technology.

General philosophical approaches are characterized by a certain level of abstraction, which often (but not always[2]) comes at the price of a marked lack of interest in technical "things", deemed to be philosophically inessential. We are reminded of Heidegger's famous phrase: the essence of technique is nothing technical; in other words, we will not gain access to what is essential in the (contemporary) technical fact by dwelling on the technical realities themselves. More recently, Dominique Janicaud, a philosopher specializing in Heidegger, made a remark in his contribution to a special issue of the *Revue internationale de philosophie* devoted to technique, "that a [...] conceptual and reflexive approach can also be a philosophy of the 'technical object' in a sense that remains philosophically unessential [...]: it shows what the technical object is, but it hardly questions the technoscientific event as such" (Janicaud 1987). Techniques themselves, their trajectory in space and time, and their link with other social realities: the philosopher preoccupied with "the essence of technique" must leave all

2 The case of Gilbert Simondon is a counterexample: Simondon never defined himself as a "philosopher of techniques", and for him, technique is a problem of general philosophy, linked to a philosophy of nature, a philosophy of knowledge, a philosophy of aesthetics and a moral philosophy. However, his approach to technique is focused on objects.

this aside. Under these conditions, it is hard to see what the philosopher could expect from the historians of techniques and vice versa. On the contrary, the institutionalization of the philosophy of technology as a specialized discipline in the 1970s went hand in hand with a refocusing of analyses on the specific, sociohistorical aspects of the technical fact – a so-called "empirical turn" in the philosophy of technology (Kroes and Meijers 2000; Achterhuis 2001; Brey 2010). This was to lead (and indeed has led) philosophers of technology to allow themselves to be taught by the empirical disciplines of the humanities and social sciences. Does this mean that, since the 1970s, the philosophy of technology has been naturally open to dialogue with representatives of these disciplines, particularly with historians of techniques? It must be said that this is not quite the case.

This chapter begins with a contrasting diagnosis of the actual state of dialogue and collaboration between the philosophy of technology and the history of techniques. Second, it touches on the observation that the relationships between the two disciplines are rather distant and evokes the long-standing existence of fruitful links between them from different perspectives. Ultimately, the chapter concludes by inviting us to consider the interdisciplinary work of philosophers of technology and historians of techniques as a task that, for the most part, remains to be accomplished.

As the author of this chapter is a philosopher, not a historian, the relationship between the philosophy of technology and the history of techniques will be considered from the point of view of the former. It will become clear that philosophers of technology benefit from situating not only their objects of reflection but also their concepts and problems in a historical perspective. They therefore absolutely need historians. However, is the opposite true? It is up to historians of techniques to answer this question and to decide how fruitful it would be, from their point of view, to strengthen collaboration with philosophers of technology.

11.2. The beginnings of a dialogue between two young disciplines, from the 1970s onward

In a 1995 article, the philosopher Joseph C. Pitt confided that several years earlier, during a conference on the philosophy of technology and the history of techniques held at Eindhoven University of Technology, he had suggested to a distinguished fellow historian of techniques that the Society for the History of Technology and the Society for Philosophy and Technology (the learned society for the philosophy of technology, of which Pitt was then vice-president) should organize a major scientific event bringing the two communities together. The response from Pitt's colleague was scathing: "Oh no! The SPT hates technique. Furthermore, they know nothing about it. What could we possibly have to say to them?"

(Pitt 1995). This reaction speaks volumes about the opinion that historians of techniques seemed to have of philosophers, who claimed to talk about technique.

The Eindhoven conference was not the first to bring together historians and philosophers of technology. A similar conference had already been held at the University of Illinois in 1973. The proceedings were published six years later (Bugliarello and Donner 1979). The table of contents of this volume is instructive: the contributions of philosophers and historians are brought together in two distinct parts, entitled *History of Technology* and *Philosophy of Technology,* respectively, with the third and final parts of the work laconically entitled *The Future of Technology*. A juxtaposition of disciplines, rather than a genuine dialogue between them, is the impression given by this volume. The review written by historian Edwin T. Layton Jr. for the *Isis* journal in 1980 is highly critical (Layton 1980). The work is considered rather poor – especially the philosophical part. It is a harsh judgment that amounts to saying the same thing as Pitt's colleague a decade later: in regard to technique, philosophers have little of interest to say. Out of the entire philosophical section of the collective volume, there are only two favorable contributions in Layton Jr.'s eyes, including Mitcham's, which we will have to return to. Even the more positive reviews struggle to identify the foundations of a constructive dialogue between the two disciplines. After noting the book's lack of a guiding theme, philosopher Bertram Morris noted in his review that what all the contributions have in common is that they take a humanist view of techniques. This is not worth much.

The historians' unappealable judgment of the philosophers, which was based on a conference held in 1973, is understandable at a time when philosophical reflection on techniques was still largely the work of those who would later be called the "classical" philosophers of technology: the existentialists and phenomenologists (Heidegger, Jaspers), the Marxist-inspired Frankfurt thinkers (Marcuse, Habermas) or the Christian thinkers (Ellul, Berdiaev), whose strong point was certainly neither the fine description of techniques in context nor a concern for historical detail.

It is true that by the early 1970s, the landscape for the philosophy of technology was already far more varied than that. In the first anthology on the philosophy of technology published by Mitcham and Mackey in 1972 (Mitcham and Mackey 1983), for example, we not only find contributions from the "classical" thinkers of technology[3] (Ellul, Anders, Berdiaev, Jünger, Ortega y Gasset and Jonas) but also contributions from philosophers proposing a new line of questioning from

3 Some important names are missing, such as Arendt, Marcuse and Heidegger, but the editors of the anthology explain: their absence is not an oversight, but a deliberate omission for a variety of reasons – problems of translation rights concerning Heidegger, for example.

the "classical" approaches, including those of Mitcham himself, as well as Mario Bunge, James Kern Feibleman and Henryk Skolimowski, among others. Some of them were present at the 1973 conference – Mitcham, Bunge, Skolimowski, to which other names were added, testifying to a diversification of approaches.

However, in the early 1970s, philosophical reflection on techniques continued to offer historians of techniques few focal points. This is clearly reflected in the general position of the 1972 anthology. In their introduction, Mitcham and Mackey set out to identify the specific aim of the philosophy of technology, justifying its existence as a discipline in its own right. According to them, it is distinguished, above all, by the type of questions it poses to techniques, which are different from those of interest to the engineer or technician. The coeditors mention that a good way of distinguishing between technological and philosophical questions would be to say that the former, unlike the latter, are limited to concrete issues. The philosophy of technology, on the other hand, relies much more on reasoning than on empirical information. Therefore, what kind of dialogue is possible between philosophers and historians of techniques? The anthology does include contributions from two historians, Lynn White Jr. and R.A. Buchanan, but these are speculative contributions to the relationship between techniques and the Christian religion – Mitcham and Mackey are not saying that only professional philosophers are entitled to philosophize! Historians can too, of course. However, this still does not lead to an interdisciplinary dialogue between historians and philosophers of technology, which, of course, was not the aim of the 1972 anthology.

From this point of view, Mitcham's contribution to the 1973 conference and the 1979 volume proceedings merits particular scrutiny, since it confirms (without intending to) the observation of a gap between the two scientific communities (Mitcham 1979). Mitcham begins by distinguishing three ways in which philosophers of technology relate to history: first, there are "analytical style" philosophers[4], who look to history for examples to confirm their conceptual analyses; here, history simply provides illustrations to support philosophical theses. Next are philosophers who view the succession of what we might call the "eras" of techniques, each embodied by an emblematic invention (the steam engine for the industrial "era", etc.), as the manifestation of a global meaning, as is the case with philosophies that see technical "progress" as the driving force behind human progress. Here, history is the general framework within which techniques, as a global process, impact human beings. Finally, there are philosophies that match the prevailing conception of techniques in a given society with general positions of a

4 In a sense that is not what is usually called "analytic philosophy".

metaphysical, ethical or political nature. Mitcham associated these three ways of philosophizing about technique with three ways of writing the history of techniques.

The first, corresponding to the "analytical" style in philosophy, consists of writing a technical history of techniques, focusing on how objects are made and used. Mitcham noted that this type of history of techniques is simpler to write than other histories, such as those of the arts or sciences, in that it consists first and foremost of a kind of chronology of inventions. It is simpler because it is strictly factual, whereas the history of the arts, for example, inevitably involves a certain amount of interpretation on the part of the historian. Let us note to what extent here Mitcham seems far, very far even, from the way historians view the technical history of techniques. In a short seminal article that was the first of its kind in France, Lucien Febvre also maintains that a history of techniques is, first and foremost, a technical history of techniques, but he certainly does not consider it a chronology of inventions, according to a linear and oriented arrow of time. Instead, he advocates following what he calls the "journey" of techniques in space and time, which is much better suited to making us aware of diversions, stops and backtracks, in contrast to any idea of linear succession (Febvre 1935).

The second way of writing the history of techniques that Mitcham speaks of, corresponding to philosophies that view technique as the embodiment of a global meaning, is to study techniques in terms of their overall impact on societies and human beings. In this category, we can find figures as diverse as Marx, Heidegger and Marcuse. Here, the philosophy of technology has touched upon the thesis of "technological determinism"; in other words, the thesis that claims technique causally determines the characteristics of the societies in which it is used. Of course, things are much more complex than that, and it would be inaccurate to call Marx univocally "deterministic" and even more so for Heidegger and Ellul. In any case, these thinkers consider techniques from perspectives that seem frankly incompatible with the methodological requirements and goals of historians. "A philosopher's work, much more than a historian's", says Febvre of Cournot. L. Mumford, often cited as a historian and judged very harshly by Daumas: to categorize history into great periods defined by the association between materials and a preferred energy source, as Mumford maintains, is certainly not to satisfy the requirements of scientificity that historians of techniques advocate for in their discipline (Daumas 1969).

The third way of writing the history of techniques, again according to Mitcham, is to propose a history of ideas about technique – how has human industrial activity been thought about and assessed through the ages? This historical perspective corresponds to the third philosophical posture Mitcham mentions – the least developed, in his view. In the remainder of his contribution, he proposes an illustration of the results that this third type of entanglement between philosophy and the history of techniques can produce. To this end, he examines the Greek notion of *techne* in great detail and then compares it with the modern notion of technique, pointing out the fundamental differences between the two notions – differences that essentially involve the conception of matter in the ancient Greek and modern worlds, respectively. For all the philosophical interest and finesse of Mitcham's analysis of Plato in particular, it is hard to see how the historian of techniques could find some sort of historical view of techniques that makes sense to him/her. Incidentally, Mitcham himself does not seek to defend the idea of fruitful dialogue between two established scientific disciplines (in the process of being institutionally constituted, as far as the philosophy of technology is concerned). His aim is different; rather, it is to open up a new territory for philosophers who are interested in techniques, a territory constituted not by techniques themselves but by the ideas we have historically formed of them. A philosopher's history, not a historian's history, over and over again.

Mitcham's emphasis on the history of ideas about technique is echoed in Elisabeth Ströker's contribution to the proceedings of a conference held in 1981 in West Germany, which brought together North American and German philosophers of technology (Durbin and Rapp 1983). Ströker noted that the philosophy of technology, a young discipline, is still struggling to clarify its concepts and systematize itself in the same way as other, more mature philosophical disciplines. To overcome this situation, Ströker advocates placing the problems presented by philosophers of technology in the context of a long history, particularly those concerning the relationship between science and technique – problems that are entwined and difficult to untangle if approached in synchronicity, in a short time frame. The concepts used by philosophers of technology in their analyses also benefit from being placed in the history of their formation and transformation over time. The philosophy of technology therefore needs to be historicized by assuming that its problems, the way it presents them and the concepts it mobilizes, are historical realities. This excellent advice, which has a certain kinship with the philosophies of technology inspired by the "French style" method of historical epistemology, still does not lead to an actual dialogue between philosophers of technology and historians of techniques.

In summary, the 1970s and early 1980s were undeniably marked by the development of a new style of philosophy of technology, but this did not yet bring about a truly fruitful dialogue between philosophers of technology and historians of techniques, despite some attempts to bring the two communities closer together. Notably, these new philosophers of technology include a number of analytical philosophers[5], such as Rapp, for whom the philosophy of technology is an a-historical discipline. Thus, in his introduction to the anthology of the analytic philosophy of technology he published in 1974, Rapp explicitly assumes that the contributions to this anthology take no account of the historical development of techniques (Rapp 1974).

11.3. The "empirical turn" in the philosophy of technology and its impact on the relations between philosophers and historians of techniques

Although Layton Jr.'s harsh judgment was understandable in 1980, the reaction of Pitt's colleague in the late 1980s was even more unexpected. Indeed, the creation of the SPT in the mid-1970s went hand in hand with the emergence of a new generation of philosophers for whom technique was, as it had been for their elders, a philosophical object of the utmost importance, but who, unlike the "classical" philosophers, opted for a more empirical, contextual and nuanced approach to contemporary technique. At the end of the 1990s, the next generation of philosophers of technology meaningfully dubbed the revival of philosophical reflection on techniques in the late 1970s and 1980s the "empirical turn". An "empirical turn" means a greater focus on techniques in their concrete aspect, whether this involves examining engineering activities and innovation processes and, in particular, the type of knowledge used in the latter; the relationships between science and technique, which are more complex and entangled than the definition of technique as "applied science" suggests; the way in which the uses of techniques transform human existence in society, as well as social interactions; the ethical and political aspects of techniques; and the specificity of technical artifacts in comparison with other types of artifacts, all of which are likely to encourage a closer link between philosophers of technology and historians of techniques, insofar as these new-style philosophical analyses were (and still are) largely based on the examination of cases and examples taken from more or less recent history. In a volume entitled *A Companion to the Philosophy of Technology* (Misa 2009), the historian of techniques Thomas J. Misa emphasized in the late 2000s the importance of the "empirical turn" in the philosophy of technology for the establishment of a fruitful dialogue between philosophers and historians. He noted that, prior to this "turning point", there was no possible dialogue between the two scientific communities. On the question of technological determinism, you could not find anyone more opposed than a philosopher and a historian: "classical" philosophers,

5 Not in the sense given to the term by Mitcham in his typology, but in the sense of the neopositivism that dominated the philosophy of science in the English-speaking world at the time.

according to Misa, were inclined to consider, on the one hand, that technique determines social and cultural change and, on the other hand, that it develops autonomously – two assertions that a historian could only categorically reject. Misa emphasized that the "empirical turn" has reshuffled the cards and led philosophers of technology and historians of techniques to discover common interests and shared questions. *A Companion to the Philosophy of Technology* is divided into seven parts, the first of which is *History of Technology*; *Technology and Philosophy* (not *Philosophy of Technology*) constitute the third, behind a section devoted to examining the relationship between science and technique; and most parts are devoted to sectoral issues: technique and the environment, technique and politics, technique and ethics, plus a final section on technique and the future. In short, compared with the 1979 volume, the 2009 book demonstrates that philosophers and historians are indeed now able, and willing, to engage in dialogue on specific themes. By the end of the 2000s, philosophers of technology now exist as a well-established international community, no longer having to worry about justifying their existence as such, which was not the case in the late 1970s. In *A Companion to the Philosophy of Technology*, the aim is not to establish a new discipline and defend its legitimacy but to bring it into conversation with other disciplines, primarily, in this case, the history of techniques.

The fact that the historian of techniques Pitt approached at the Eindhoven conference gave him such a scathing reply at the end of the 1980s, when the "empirical turn" was already well established and had already produced important works by the likes of Don Idhe, Andrew Feenberg and Mitcham, is perplexing. It should be noted, however, that Misa dates the "empirical turn" back some ten years before the publication of *A Companion to the Philosophy of Technology*, in other words, to the late 1990s, which is inaccurate since it should be remembered that the expression "empirical turn", which actually appeared in the late 1990s, refers to a new style of philosophical thought that emerged some 20 years earlier. Misa is, in fact, talking about what we might call a second "empirical turn", which effectively marked the 2000s and to which we will have to return. Should we assume, then, that the interdisciplinary dialogue between philosophers of technology and historians of techniques only began at the very end of the 20th century, during this second "empirical turn", but not the first? Certainly, the "empirical" repositioning of the philosophy of technology in the 1970s took place much more in dialogue with science and technology studies than with the historians of techniques. Important references here are both constructivism (social construction of technology) and actor–network theory. Historians, such as Steven Shapin and Simon Schaffer, are certainly among these important references but, on the whole, the opening of the philosophy of technology to empirical disciplines was not massively directed toward the history of techniques, at least if we are to believe Misa, until the very end of the 1990s, when things really began to change.

However, this diagnosis is not entirely accurate. Following the 1981 conference in West Germany, a second conference was held in 1983 in New York, this time on a specific theme: information technology. The proceedings were published in 1986 (Mitcham and Huning 1986). In their preface, the two editors, Mitcham and Huning, begin by welcoming the recent cooperation between the philosophy of technology and the history of techniques, following the cooperation that had been taking place for two or three decades between the philosophy of science and the history of science. More precisely, while the two prefatory authors emphasize the valuable contribution of the history of science to the philosophy of science, in the case of techniques, they stress the fruitfulness of the contribution in the other direction – from the philosophy of technology to the history of techniques: traditional historical studies of techniques, they note, are now supplemented by philosophical questioning, and no longer focus solely on the contexts of the application of techniques, on invention and engineering, or on the reciprocal relationships between technique and society. This is a curious way of looking at things, given that the "empirical turn" in the philosophy of technology went hand in hand with the interest that philosophers began to show in, among other things, the processes of innovation, engineering and engineering knowledge, as well as the social contexts of innovation and the use of techniques. As early as 1983, Mitcham and Huning stressed the importance of interdisciplinary dialogue between philosophers of technology and historians of techniques. The fact that a historian of techniques should be so fiercely opposed to the idea of co-organizing a conference with a philosopher in the late 1980s, as evidenced by the historian's rather blunt response to Pitt, seems somewhat excessive and even unfair. Undoubtedly, bridges had already been built between the two scientific communities.

The desire to collaborate became increasingly clear in the 1990s, especially at the turn of the new century. Misa is right to emphasize this point when, in 2009, he referred to the links highlighted over the last 10 years or so. However, in addition to the conferences already mentioned in 1973 at the University of Illinois and in 1983 in New York, and even before we reach the very end of the century, we must mention a major meeting between philosophers of technology and historians of techniques in 1990 – a meeting organized around the theme of the relationship between science and technique, and explicitly motivated by the need, to properly address this complex issue, to bring the two disciplines into dialogue. While the 1973 conference, and the publication of the proceedings in 1979, did little to create the conditions for such a dialogue and while the 1983 conference, and the publication of the proceedings in 1986, gave a glimpse – but only a glimpse – of the benefits of a greater link between historical studies and philosophical questioning, it was truly only in 1990, during this meeting on the problem of the relations between science and technique, that the need for an interdisciplinary dialogue associating philosophers of technology and historians of techniques was forcefully asserted. The proceedings of this meeting, published in 1992 by Peter Kroes and Martijn Bakker, leave no room for ambiguity (Kroes and Bakker 1992). The introduction to the volume is essentially devoted to demonstrating that neither historians nor philosophers alone, both carefully avoiding talking to one another, will succeed in clarifying the terms of the debate on the relationship between science and technique. To do so, we first had to shift these very terms to present the problem from a new perspective: not that of the relationship between two supposedly defined activities (science and technique) but that of the type of knowledge involved in the activities of engineers and technicians. The meeting thus concerns "technological knowledge", which can be satisfactorily understood only by explaining its historical context of implementation, on the one hand, and the epistemic specificities that distinguish it from scientific knowledge, on the other hand. Naturally, historians and philosophers must work together, as the phenomenon under study demands it. From 1973 to 1990, we have come a long way. This is clear from the table of contents: while the proceedings of the 1973 conference give the impression of a juxtaposition of perspectives rather than a genuine dialogue, this is no longer the case in the proceedings of the 1990 meeting; contributions from philosophers and historians alternate, in equal numbers or more or less; there is certainly no question of texts cowritten by a historian and a philosopher, but the unity of the volume is much more assertive.

Less than a decade later, in 1999, another conference was held, affirming the need for interdisciplinary dialogue between philosophy and the human and social sciences. The aim of this conference was to examine the concepts of modernity and technology together. The initial observation is that specialists in the first concept (philosophers, social theorists) generally have a very abstract conception of technique whereas, on the contrary, a strong sensitivity to the diversity of techniques and their manifestations is needed to correctly apprehend the phenomenon of "modernity"; specialists in technology studies, however, do consider the diversity of technical facts, which are protean and omnipresent but, more often than not, they merely accumulate empirical analyses without any ambition to organize them according to a global view. Here, again, the organizers of the conference are driven by the conviction that philosophers and specialists in empirical disciplines – not just technical historians, but also sociologists and anthropologists – must work closely together. The proceedings of this conference, published in 2003, forcefully make this point (Misa et al. 2003). Notably, for the first time, these proceedings are codirected by a historian (Misa) and two philosophers (Brey and Feenberg).

Things were therefore well underway in the 1990s. One detail, however, is that the 1990 meeting took place at the Eindhoven University of Technology. Was it not the same place, a few years before he published his 1995 article, that Pitt received a scathing reply from his fellow historian? There is no denying that interdisciplinarity is increasingly being promoted on the editorial front, in prefaces and in introductions to the proceedings of successive conferences, but this motto may not yet have found its way into all minds – at least those of historians of techniques. Philosophers have yet to convince their colleagues that they have anything valid to say about techniques.

Have philosophers fully grasped the historical nature of their concepts and the problems they pose, as Ströker was already urging them to do in the early 1980s? We cannot be sure. In 2003, on the occasion of the publication of a new anthology on the philosophy of technology (Scharff and Dusek 2003), Don Ihde wrote a review and took the opportunity to take stock of the young discipline after some 30 years of existence[6] (Don Ihde 2004). At the very end of his text, after the diagnosis in which he suggested a few pointers for orientation, Don Ihde mentioned two subjects that could usefully bring the philosopher closer to the historian.

The first topic concerns “imagined technologies”, in other words, technologies that have been thought out, described and designed, such as Leonardo da Vinci’s machines. These “imagined technologies” carry a great deal of weight in contemporary imaginations and in the anticipatory speeches that accompany major research and development programs based on the naïve conviction that they will one day become reality. Here, teaching technical history can prove invaluable in revealing that, contrary to this naïve conviction, many of these “imagined technologies” were never actually manufactured and put into operation. The message here is that these imaginary concepts should not be taken at face value.

According to Don Ihde, there is a second subject that should encourage philosophers of technology to learn from historians of techniques: technical metaphors. We know, for example, the importance of the clock as a model of the universe in classical thought. Would the philosopher not benefit from being taught by a historian who could show them the evolution of clockmaking techniques and the importance of mechanical timekeeping in medieval and modern societies?

While Don Ihde’s two pieces of advice are undeniably relevant, we must admit that there is little space for dialogue between the philosophy of technology and the history of techniques here and that it is limited to very specific motivations. We are

6 If we trace the birth of the philosophy of technology itself back to the early 1970s, this is not a consensual point.

a long way from the great calls for interdisciplinary collaboration that motivated the conferences of the 1990s. Admittedly, the volume reviewed in Don Ihde's text is a disciplinary anthology, so it is understandable that the relations of the said discipline, the philosophy of technology, with other disciplines should not have been at the heart of this review. Nevertheless, given that Don Ihde has given this review a much broader scope than that of a simple review, the fact that the dialogue between philosophers and historians appears rather inessential tells us that, at the beginning of the 21st century, the great calls of the 1990s are still waiting to be implemented. It seems that if there is dialogue, it is superficial.

Let us return to Misa's encouraging diagnosis at the end of the 2000s: the "empirical turn" (the second turn, the one that marked the 2000s and beyond) would therefore have given a decisive push to the link between the philosophy of technology and the history of techniques. However, what actually happened? This second "empirical turn" is characterized by three new orientations in the philosophy of technology: first, a focus on "technical artifacts" from two additional perspectives, which are post-phenomenology, as represented by Don Ihde and Peter-Paul Verbeek, and analytic philosophy, as represented by Pieter Vermaas, Peter Kroes, Anthony Meijers, Maarten Franssen and Wybo Houkes, among others; second, the growing importance of ethics; third, the evolution of philosophers of technology toward a more interventionist stance – leading them to make design a priority issue.

The post-phenomenological trend focuses on philosophical analysis of "technological mediations" and their design (Verbeek 2005). The analytic trend, however, describes and characterizes "The Dual Nature of Technical Artifacts", in other words, both their material and intentional natures: every technical artifact results from the combination of these two natures (Kroes and Meijers 2006). What is the place of the history of techniques in these two research programs, and more generally, in the new orientations of the philosophy of technology? It has to be said that it is not central. It is non-existent in the analytical program The Dual Nature of Technical Artifacts, since analytical approaches in philosophy are largely ahistorical. Nor is it very present in Verbeek's post-phenomenological program, which is much more concerned with pragmatism, STS and, above all, actor–network theory than with the history of techniques.

Contributions to the ethics of techniques, on the other hand, are more concerned with making normative proposals than with situating problems in a historical perspective.

Finally, philosophical interventionism has no use for the historian's teachings (except perhaps to compare current forms of philosophical intervention with, for example, the more classical figure of the committed intellectual, but this is entirely anecdotal).

Does Misa's diagnostic approach not seem over-optimistic? Beyond the repeated calls for dialogue and the summaries of collective works striving to interweave the contributions of philosophers and historians, what is the reality of interdisciplinary work? If we look no further than the editorial benefits of the conferences devoted to dialogue between philosophers of technology and historians of techniques over the past fifty years[7], the picture looks rather bleak.

The institutionalization of philosophy of technology as a specialized discipline since the 1970s does not seem to have stimulated any real interdisciplinarity between the philosophy of technology and the history of techniques, at least in terms of learned societies, major scientific events calling for collaboration, and their editorial benefits – despite the context of the "empirical turn".

What happens to this interdisciplinary dialogue when we consider the philosophy of technology as a general philosophy rather than as a specialized discipline? We must admit that the situation seems just as complicated.

11.4. Before the 1970s

Throughout the 20th century, there have undeniably been efforts to clarify the history of techniques and philosophical reflections on techniques, embodied by a number of personalities with singular trajectories. In the end, however, these efforts seem to have ultimately resulted in a widening, rather than bridging, of the gap between the two perspectives. There are several examples of this in France.

Let us start at the beginning with Lucien Febvre, credited with co-founding the "Histoire des techniques" ("History of techniques") discipline, along with Marc Bloch, in the aforementioned seminal article (Febvre 1935). In this context, Febvre structures the new discipline in three chapters: first, the technical history of the techniques mentioned above; second, a history interested in identifying the driving forces of technical change between theory and practice, with a focus on the relationship between science and technique; and last, a history placing technique in

7 That is, from the moment when the latest of these disciplines, the philosophy of technology, entered the institutional sphere (through the creation of a learned society, a journal and the holding of dedicated conferences).

the context of all human activities – religion, art, politics, etc. – with a view to understanding how the history of techniques affects general history and vice versa. This program is marked by Febvre's career and, in particular, by his long-standing relationship with the philosopher Henri Berr (1863–1954). In his doctoral thesis, defended in 1898, then with the creation of the *Revue de synthèse historique* in 1900, the launch of *L'Évolution de l'humanité*, and the creation of the Centre international de synthèse in the late 1920s, Berr tirelessly sought to defend two major ideas: first, history must become a science on par with the natural sciences, but one that satisfies its own criteria of scientificity (there is therefore an epistemology specific to history); second, if knowledge testifies to the progress of reason, and as such must be integrated into a vast synthesis – an idea that was not original at the time – then this is no longer the domain of the philosopher, but of the historian – an original idea, on the contrary, that distances Berr from philosophies of history. Febvre became acquainted with Berr's work in the early 1910s and was enthusiastic about his concept of a non-event-based history, which did not consist of an alignment of facts and dates, as opposed to the 19th-century historians such as Seignobos and Langlois, from whom Febvre was trying to distance himself. Febvre was also receptive to the idea of a link between history and the other human and social sciences (Durkheimian sociology for Berr, geography and economics for Febvre). All of these factors brought the two men, the philosopher and historian, closer together, at least until the 1930s. From then on, relations between Febvre and Berr became strained. For epistemological reasons, let us say that Berr seems to have remained too much of a philosopher in Febvre's eyes, too much of a positivist in his conception of historical synthesis, whereas Febvre only conceived of history as emptied of any excessively speculative or even metaphysical notion (Le Roux 2000). Febvre's 1935 article bears the mark of this refusal. A dig at Cournot is evidence of this. Referring to Cournot's reflections on the relationship between science and technique, Febvre writes, "the work of a philosopher, much more than that of a historian". However, Febvre's emphasis on interdisciplinary collaboration reflects Berr's idea of historical synthesis.

In short, the founding article of the "history of techniques" discipline is intended to circumscribe its territory, not only in opposition to the philosophies of history that were the heyday of the 19th century but also by distancing the philosopher from the subject that seems to concern him most: the relationship between theory and practice, between science and technique (Chapter II of the history of techniques according to Febvre). According to Febvre, philosophers, even the best ones such as Cournot, can propose only speculative discussions on this subject rather than a documented analysis.

Another observational trajectory illustrates even more clearly that the history of techniques does not seem to mix well with philosophical speculation: that of the Jesuit François Russo. As Ronan Le Roux noted, from the end of the 1940s to the publication of his *Introduction à l'histoire des techniques (Introduction to the history of techniques)* in 1986, Russo wrote approximately 70 texts devoted to technique, including philosophical reflections, analyses of Catholicism's position on technique, and texts on the history of techniques itself (Le Roux 2013). It would be a mistake, however, to believe that Russo wanted to perform several separate tasks at once; a link exists between these different types of contributions. Russo's ambition was indeed to provide a general conception of human action on matter, whether it was mediated by machines or not. According to Russo, the cybernetic concept of information was intended to provide a unified theory of technical operations. With the theory of informational machines, Russo aimed to go beyond the conception of the history of techniques as an accumulation of facts and dates and to identify an overall structure and logic of technical evolution (in its relationship with science). This evolutionary conception of the history of techniques led Russo to embrace the views of Teilhard de Chardin, who saw technical evolution as a kind of journey of the human species toward fulfillment. Through technical evolutionism, theological and philosophical reflections are thus interwoven with historical analyses. However, this evolutionary perspective is no longer apparent in *Introduction to the History of Techniques*. Russo seems to have been led to renounce it. Le Roux explains this: in the 1980s, "the history [of techniques] became professionalized in order to detach itself from the 'great systems' of interpretation, from global visions of the world, no doubt to the point of being able to oppose a robust epistemological barrier to such strategies". For the history of techniques, becoming more professional inevitably meant abandoning philosophical notions.

Should we confine ourselves to this general observation that the dialogue between the history of techniques and the philosophy of technology is difficult, to say the least, or even lacking overall? Absolutely not. This interdisciplinary dialogue has indeed been established, but for the most part elsewhere than in the meetings, which, paradoxically, have been promoting it for some 40 years. In fact, it is easily perceptible from four perspectives: that of individual trajectories; that of traditions of thought, some of which are much more inclined than others to interweave philosophy and history in their study of techniques; that of shared concepts; and that of a historicization of the concepts and issues of the philosophy of technology.

11.5. Concrete elements of interdisciplinary dialogue

Let us look at individual trajectories first, while limiting ourselves to a few emblematic examples. In addition to Russo, who nevertheless endeavored to combine historical studies with philosophical reflections, we should mention Pierre Ducassé. A philosopher and historian of techniques, Ducassé, served as a general secretary for scientific works at the Institut d'histoire des sciences et des techniques (French Institute for the History of Science and Technology), from its creation in 1932 until the arrival of Canguilhem, who took over as director in 1955 (Gayon 2016). After writing his thesis on Comte, Ducassé distanced himself from positivism, rejecting the idea that the development of techniques could be organized and guided by theory. To think philosophically about technique, we must start from its concrete historical manifestations (Le Roux 2011). For Ducassé, as for Russo, the aim is to identify a general logic provided by the concept of technical evolution.

More recently, the works of Hélène Vérin and Ann Johnson are also good examples of this close interweaving of epistemological reflection and historical investigation. For example, Johnson's study of a braking safety system for motor vehicles is a combination of both the social history of the engineering communities involved and an epistemological analysis of engineering knowledge (Johnson 2009). Johnson died in 2016 at the age of 51 and was honored by both the Society for Philosophy and Technology and the Society for the History of Technology at their respective conferences in 2017, testifying to the recognition of her work as both a historian and an epistemological philosopher.

Bernadette Bensauve-Vincent has also done much to promote dialogue between philosophers and historians of science and techniques (and, beyond that, with STS). Rewarded with the Sarton Medal in 2021, reflecting international recognition for her work in the history of science, Bensaude-Vincent is nonetheless a philosopher in her own right, for whom contemporary technical facts lead to the reformulation of ancient philosophical issues, such as the relationships between form and matter, between the natural and the artificial, between science and technique, and between science and opinion. The collective publications she has coedited show how there is a constant concern with cross-disciplinary perspectives on themes that require, precisely, this cross-fertilization (Bensaude-Vincent and Newman 2007; Bensaude-Vincent et al. 2017, 2022; Bensaude-Vincent 2022). Materials (from chemistry, nanotechnology), man-made living beings from synthetic biology, plastics and nuclear power: all of these objects appear less as entities stabilized by their *eidos*, in other words, by an essential and immutable definition, than as uncertain, labile, multiscale trajectories, tying disparate realities together and tirelessly undoing these links to tie others together (microphysical processes, metabolisms, values and interests and heterogeneous temporalities). The mode of existence of these often bizarre and puzzling objects, for which philosophical tradition has left us no satisfactory concept, can be genuinely studied only by crossing the perspectives of the philosopher, the historian, the anthropologist and the STS.

Ian Hacking, Arnold Pacey, Walter Vincenti, Davis Baird and Adelheid Voskuhl could also be cited, and the list is by no means exhaustive. There should also be mention of the increasingly obvious propensity of the younger generation, those now working on their theses, to closely articulate philosophical questioning and the historical approach, in other words, to construct the philosophical object as a sociohistorical object, without worrying too much about disciplinary boundaries – a very beneficial freedom that should be welcomed.

In terms of traditions of thought, while the post-phenomenological and analytic approaches in the philosophy of technology do not seem inclined (or very modestly so) to integrate the history of techniques into their analyses, the epistemological history of the sciences approach is, on the contrary, naturally inclined to do so. While Canguilhem is of course a major reference, particularly for the IHST seminar held between 1963 and 1965, in which Simondon, Jacques Guillerme and Jan Sebestik took part – a seminar from which the latter two drew their article "*Les commencements de la technologie*" (The beginnings of technology), published in Thalès in 1968 – we must also mention the thinking of François Dagognet, whose *Éloge de l'objet* combines philosophical considerations with a historian's erudition on a number of techniques (glass and chairs, for example). More recently, the so-called "biographical" approach to technical objects, popularized in material anthropology and now taken up by Bensaude-Vincent, has also led to the link between philosophy and history. Importing the biographical method into the philosophical study of techniques is justified by the idea that a technical object is a weaving of entangled histories, unfolding in multiple and heterogeneous temporalities, inextricably combining natural operations and human activities (Bensaude-Vincent 2012). Unlike analytical discourse, narrative enables us to hold these heterogeneous histories together and understand the object as a bundle of meanings. Thus, according to Bensaude-Vincent, the biographical approach makes it possible to "reveal all that is concealed, hidden behind the practice of scientific and technical reason" (Bensaude-Vincent 2012).

Philosophers and historians have also established fruitful dialogue around certain concepts. The aforementioned concept of technical evolution is a good example. In the case of Russo, this concept may seem incompatible with the methodological requirements of the history of techniques. However, the situation is far more complex. The concept of technical evolution has been used in archaeology, anthropology and paleoanthropology, as well as in machine science, since the 19th century. In the history of techniques, Maurice Daumas also found in the concept of technical evolution a way to go beyond the conception of his discipline as a chronology of inventions, following the path opened up by Febvre. Daumas explicitly acknowledged that this concept had come to him from reading Maunoury, who himself was a reader of Simondon. As a matter of fact, Simondon's commentators agree that evolution and history are two different perspectives on the technical object that should not be confused. More recently, the concept of technical evolution has been criticized on the grounds that it suggests a linear, deterministic conception of technical change, without consideration for the social determinants, contextual specificities and divisions that make up the reality of the history of techniques. However, a historian of technology such as George Basalla has made it the guiding concept of his work. This concept of technical evolution is undoubtedly one of the most delicate and difficult to define in the interdisciplinary dialogue between historians of techniques and philosophers of technology.

Other concepts can give rise to fruitful exchanges between historians of techniques and philosophers of technology, such as analogy and technology. The concept of analogy is emblematic of these interdisciplinary clashes. On the philosophical side, analogy is a major concept in Simondon's analysis of techniques, through the idea that they must be studied in terms of the operations they underlie, and not as a separate domain, separated from other domains such as biological or social facts, each with its own characteristics – in the case of technique, this would be usage, the fact of being in view of, of serving, of being used for such and such a specific purpose. As soon as techniques are analyzed through the prism of their operations, rather than their structures and purposes of use, they become accessible for comparative study with domains that are phenomenally very different (the living, the social, for example) but that may specifically present operative analogies with them. Analogy is then the method of general technology, which Simondon conceived as a science of operations in the Being. In the technical field, the concept of analogy leads us to question the cognitive processes of invention – a perspective that resonates with the program of a technical history that is focused on the processes of intellection at work in all technical activities and the complexity of technical intelligence. The concept of analogy has thus led historians, cognitive scientists and philosophers of cognition to compare their perspectives, the former being responsive to the numerous uses of analogy and meanings of the term throughout history and the latter being more inclined to establish similarities in these uses and to identify the general motives of reasoning by analogy (Archambault de Beaune et al. 2017).

The concept of artifact, which makes sense for historians and is at the heart of current philosophical reflections in the philosophy of technology, can also give rise to equally fruitful exchanges. A session at the 2021 SPT conference on the different paths of the "thingly turn", bringing together philosophers and historians of science and techniques, demonstrated the value of approaching the concept of the artifact, the technical "thing", as a concept shared by both communities, including anthropologists.

Finally, we should mention works that shed historical light on the concepts that have been used to think philosophically about technique, such as the concept of "technique" (Deldicque 2021). Here, the historical perspective allows us to understand how and why the technical fact, at a given moment, was thought through the prism of a certain concept. This inquiry is not exactly the same as that advocated by Mitcham in the late 1970s. It is more of a question of comprehending the concepts of the philosophy of technology as sociohistorical realities, from a perspective that focuses on the institutions in which they are elaborated, the professional trajectories of the individuals who handle and disseminate these concepts, the circulation of writings that mention them, and their semantic evolutions.

Ultimately, whether it is a question of constructing objects, formulating problems or forging concepts, philosophers of technology benefit from being taught by historians. In fact, history can have a very beneficial regulatory and critical function for the philosopher. A long-standing example illustrates this very well. In the middle of the last century, there was much debate about the relationship between the ancient Greeks and techniques: why did the Greeks not exploit their knowledge of machines and their operating principles for the purposes of economic production. One explanation, often put forward by philosophers such as Pierre-Maxime Schuhl, is that the Greeks had an abundance of cheap labor, in the form of slaves, and this would not have encouraged them to develop their technical knowledge for industrial production. It took the scholarly work of two historians with excellent knowledge of ancient Greece, Vernant and Gille, to put an end to this erroneous explanation. Here, the history of techniques happily corrects philosophical speculation.

11.6. Conclusion

In conclusion, while dialogue between the philosophy of technology and the history of techniques is a reality, it needs to be further encouraged at the institutional level. Initiatives already well underway include philosophers of technology and historians of techniques working together – around common concepts, specific problems (such as the relationship between science and technique, which has in the past given rise to some of the most fruitful exchanges between the two disciplines), and methods that encourage links between philosophical questioning and historical inquiry, such as the biographical method – deserve to be expanded on.

For philosophy, the importance of this dialogue with the history of techniques does not just lie in the contribution of knowledge from the latter to the former, enabling philosophical discourse to be controlled (as with Vernant and Gille's clarification of the situation of the Greeks with regard to work and machines). Nor is it simply a matter of providing the philosopher with elements of historical contextualization so that they can better think through their objects. "Replacing" techniques back into their historical context certainly helps frame philosophical reflection; however, limiting ourselves to this notion of contextualization causes a certain exteriority between the object of philosophical reflection and history.

A further step is taken when the philosopher constructs their object as intrinsically historical. Simondon had already made progress in this direction when he defined the technical object as "that of which there is genesis": it is the reality of a line of thought, unified by the coherence of a functional scheme, rather than an isolated individual. According to Simondon, the technical object, reclaimed by philosophy, is intrinsically temporal. It is not in time; it is made of time. However, asserting that the technical object is intrinsically temporal is not the same thing as asserting that it is intrinsically historical, so much so that for Simondon, the genetic perspective on the technical object is explicitly distinguished from the historical perspective. The Simondonian technical object is constituted and defined by its evolution, not by history.

Simondon was well versed in the history of techniques, as evidenced by his book *Du mode d'existence des objets techniques (The Mode of Existence of Technical Objects)* and his lectures, which contain a wealth of historical information. That said, for Simondon, if the history of techniques provides philosophical reflection with "food for thought", philosophy can conversely provide the history of techniques with criteria for judgment. The evolutionary approach, which, according to Simondon, is one of the aspects of philosophical discourse on technique, should enable the history of techniques to be assessed according to rational criteria (functional synergies, simplification of structures, integration of an environment associated with the functioning of devices, etc.). Here, philosophical discourse provides a check on historically based discourse, particularly that of progress. In this way, the evolutionary approach makes it clear that the history of the motor vehicle cannot be unambiguously regarded as technical progress. Philosophy thus has a regulatory function, not so much over the history of techniques as a discipline (which Simondon does not mention), as over discourses that extrapolate conclusions from history. However, to assign this role to philosophy, we first need to distinguish two levels in the technical object: on the one hand, an essential core made up of functional patterns, where the object is grasped in terms of its rationality; on the other hand, the layer of what Simondon calls "psychosocial overdeterminations", where the object's rationality is polluted, altered by value choices that have nothing to do with it (such as prestige or an aesthetic devoid of any functional relevance: for example, equipping cars with rear spoilers may satisfy one's aesthetic temporarily, but it is absurd from an aerodynamic point of view). The Simondonian technical object is, in short, two-layered; in other words, it maintains both its rationality and its historical/social contextualization separately.

The intensification of dialogue between the philosophy of technology and the history of techniques is therefore not only, or even perhaps primarily, an institutional issue. On the philosophical side, at least, it is primarily a question of conceptual work, requiring the philosopher to do more than "contextualize" their objects and concepts historically: to literally historicize them, that is, to consider them as made of historical material.

11.7. References

Achterhuis, H. (ed.) (2001). *American Philosophy of Technology: The Empirical Turn*. Indiana University Press, Bloomington.

Archambault de Beaune, S., Hilaire-Pérez, L., Vermeir, K. (2017). L'analogie, les techniques et les savoirs. Approches cognitives et historiques. In *L'Analogie dans les techniques*, de Beaune, S., Hilaire-Pérez, L., Vermeir, K. (eds). CNRS Éditions, Paris.

Bensaude-Vincent, B. (2012). Vies d'objets. Sur quelques usages de la biographie pour comprendre les technosciences. *Critique*, 781–782, 588–598.

Bensaude-Vincent, B. (ed.) (2022). *Between Nature and Societies. Biographies of Materials*. World Scientific Publishing, Singapore.

Bensaude-Vincent, B. and Newman, D.R. (eds) (2007). *The Artificial and the Natural. An Evolving Polarity*. MIT Press, Cambridge.

Bensaude-Vincent, B., Loeve, S., Nordmann, A., Schwartz, A. (eds) (2017). *Research Objects in Their Technological Setting*. Routledge, London.

Bensaude-Vincent, B., Boudya, S., Sato, K. (eds) (2022). *Living in a Nuclear World. From Fukushima to Hiroshima*. Routledge, London.

Brey, P. (2010). Philosophy of technology after the empirical turn. *Techné*, 14(1), 36–48.

Bugliarello, G. and Donner, D.B. (eds) (1979). *The History and Philosophy of Technology*. University of Illinois Press, Champaign.

Daumas, M. (1969). L'histoire des techniques : son objet, ses limites, ses méthodes. *Revue d'Histoire des Sciences et de leurs Applications*, 22(1), 5–32.

Deldicque, T. (2021). Canguilhem et "la technique". *Artefact*, 15, 201–238.

Durbin, P.T. and Rapp, F. (eds) (1983). *Philosophy and Technology. Boston Studies in the Philosophy of Science*. D. Reidel Publishing Company, Dordrecht/Boston.

Febvre, L. (1935). Réflexions sur l'histoire des techniques. *Annales d'histoire économique et sociale*, 7(36), 531–535.

Gayon, J. (2016). L'institut d'histoire des sciences. *Cahiers Gaston Bachelard*, 14, 15–63.

Ihde, D. (2004). Has the philosophy of technology arrived? A state-of-the-art review. *Philosophy of Science*, 71, 117–131.

Janicaud, D. (1987). Des techniques à la technoscience : l'enjeu philosophique. *Revue Internationale de Philosophie*, 161, 184–196.

Johnson, A. (2009). *Hitting the Brakes. Engineering Design and the Production of Knowledge*. Duke University Press Books, Durham.

Kroes, P. and Bakker, M. (eds) (1992). *Technological Development and Science in the Industrial Age. New Perspectives on the Science-Technology Relationship*. D. Reidel Publishing Company, Dordrecht/Boston.

Kroes, P. and Meijers, A. (eds) (2000). *The Empirical Turn in the Philosophy of Technology*. Elsevier Science Ltd, Oxford.

Kroes, P. and Meijers, A. (2006). The dual nature of technical artifacts. *Stud. Hist. Phil. Sci.*, 37, 1–4.

Layton, E.T. Jr. (1980). Book reviews: "The History and Philosophy of Technology", George Bugliarello; Dean B. Doner. *Isis*, 71(3), 505.

Le Roux, R. (2000). La correspondance de Lucien Febvre à Henri Berr. *Revue d'Histoire des Sciences Humaines*, 2, 163–168.

Le Roux, R. (2011). Pierre Ducassé et la revue Structure et Évolution des Techniques. *Documents pour l'histoire des techniques*, 20, 119–134.

Le Roux, R. (2013). Un jésuite, des machines, une histoire teilhardienne des techniques ? Russo dans le contexte de la pensée catholique française d'après-guerre. *Bulletin de la Société des Amis de la Bibliothèque et de l'Histoire de l'École Polytechnique*, 53, 17–25.

Misa, T.J. (2009). History of technology. In *A Companion to the Philosophy of Technology*, Olsen Friis, J.K.B., Pedersen, S.A., Hendricks, V.F. (eds). Blackwell Publishing Ltd, Hoboken.

Misa, T.J., Brey, P., Feenberg, A. (eds) (2003). *Modernity and Technology*. MIT Press, Cambridge.

Mitcham, C. (1979). *Philosophy and the History of Technology*. University of Illinois Press, Urbana.

Mitcham, C. (1994). *Thinking through Technology: The Path between Engineering and Philosophy*. University of Chicago Press, Chicago.

Mitcham, C. and Huning, A. (eds) (1986). *Philosophy and Technology II. Information Technologies and Computers in Theory and Practice*. D. Reidel Publishing Company, Dordrecht/Boston.

Mitcham, C. and Mackey, R. (eds) (1983). *Philosophy and Technology. Readings in the Philosophical Problems of Technology*. The Free Press, New York.

Pitt, J.C. (1995). On the philosophy of technology, past and future. *Techné: Research in Philosophy and Technology*, 1(1/2), 18–22.

Rapp, F. (ed.) (1974). *Contributions to a Philosophy of Technology. Studies in the Structure of Thinking in the Technological Sciences*. D. Reidel Publishing Company, Dordrecht/Boston.

Scharff, R.C. and Dusek, V. (2003). *Philosophy of Technology. The Technological Condition. An Anthology*. Blackwell Publishing, Hoboken.

Verbeek, P.-P. (2005). *What Things Do. Philosophical Reflections on Technology, Agency and Design*. Pennsylvania State University Press, University Park.

List of Authors

Lisa CALISTE
Inventaire général du patrimoine culturel
Montpellier
France

Guillaume CARNINO
COSTECH
Université de technologie
Compiègne
France

Delphine GARDEY
Université de Genève
Geneva
Switzerland

Xavier GUCHET
COSTECH
Université de technologie
Compiègne
France

Liliane HILAIRE-PÉREZ
EHESS
Université Paris Cité
Paris
France

Anaël MARREC
CHC, CNRS
Université Paris 1 Panthéon-Sorbonne
Paris
France

Joris MERCELIS
Heinrich Heine University Düsseldorf
Germany
and
Johns Hopkins University
Baltimore
USA

Valérie NÈGRE
IHMC
Université Paris 1 Panthéon-Sorbonne
Paris
France

Eric SCHATZBERG
Ivan Allen College of Liberal Arts
Georgia Institute of Technology
Atlanta
USA

Nathan SCHLANGER
École nationale des chartes
Paris
France

Delphine SPICQ
Collège de France-Université PSL
EHESS
Université Paris Cité
Paris
France

Catherine VERNA
ARSCAN, CNRS
UMR 7041
Université Paris 8 Vincennes-Saint-Denis
Paris
France

Index

A, B

C, D

E, F

G, H

I, J

K, L

M, N

P, Q

R, S

T, U

V, W

Printed and bound by CPI Group (UK) Ltd, Croydon, CR0 4YY
13/09/2026
14961072-0002